Application of Fermentation Technology in Animal Nutrition: 2nd Edition

Application of Fermentation Technology in Animal Nutrition: 2nd Edition

Guest Editors

Siran Wang
Lin Sun
Lei Chen
Jie Bai

MDPI

Basel • Beijing • Wuhan • Barcelona • Belgrade • Novi Sad • Cluj • Manchester

Guest Editors

Siran Wang
Institute of Animal Science
Jiangsu Academy of
Agricultural Sciences
Nanjing
China

Lin Sun
Inner Mongolia Engineering
Research Center of
Development and Utilization
of Microbial Resources
in Silage
Inner Mongolia Academy of
Agriculture and Animal
Husbandry Science
Hohhot
China

Lei Chen
College of Animal Science
Shanxi Agricultural
University
Taigu
China

Jie Bai
Pratacultural College
Gansu Agricultural
University
Lanzhou
China

Editorial Office
MDPI AG
Grosspeteranlage 5
4052 Basel, Switzerland

This is a reprint of the Special Issue, published open access by the journal *Fermentation* (ISSN 2311-5637), freely accessible at: https://www.mdpi.com/journal/fermentation/special_issues/8I06527N61.

For citation purposes, cite each article independently as indicated on the article page online and as indicated below:

Lastname, A.A.; Lastname, B.B. Article Title. *Journal Name* **Year**, *Volume Number*, Page Range.

ISBN 978-3-7258-7577-1 (Hbk)
ISBN 978-3-7258-7578-8 (PDF)
https://doi.org/10.3390/books978-3-7258-7578-8

Contents

Preface . . . . **vii**

Siran Wang, Lin Sun, Lei Chen and Jie Bai
Application of Fermentation Technology in Animal Nutrition: 2nd Edition
Reprinted from: *Fermentation* **2026**, *12*, 58, https://doi.org/10.3390/fermentation12010058 . . . . **1**

Huaxin Niu, Jayakrishnan Nair, Hee-Eun Yang, Tim A. McAllister, Eric Chevaux and Yuxi Wang
The Effects of a Heterofermentative Lactic Acid Bacterial Inoculant Containing *Lentilactobacillus hilgardii* and *Lentilactobacillus buchneri* with or Without Chitinases on the Ensiling, Aerobic Stability, and In Vitro Ruminal Fermentation of Whole-Crop Corn Silages
Reprinted from: *Fermentation* **2026**, *12*, 29, https://doi.org/10.3390/fermentation12010029 . . . . **6**

Yara América da Silva, Marco Antonio Previdelli Orrico Junior, Marciana Retore, Gessí Ceccon, Isabele Paola de Oliveira Amaral, Ana Carolina Amorim Orrico, et al.
Productivity, Fermentation Parameters, and Chemical Composition of Silages from Biomass Sorghum Hybrids in Ratoon Crop
Reprinted from: *Fermentation* **2025**, *11*, 540, https://doi.org/10.3390/fermentation11090540 . . . **26**

Long Zhang, Shiqin Liu, Xuepeng Wang, He Wang, Songze Li, Yuguo Zhen and Xuefeng Zhang
Effect of Harvesting Time on Starch Degradation in Rumen of Whole-Plant Corn and Its Silage
Reprinted from: *Fermentation* **2025**, *11*, 522, https://doi.org/10.3390/fermentation11090522 . . . **41**

Zhumei Du, Ying Meng, Yifan Chen, Shaojuan Cui, Siran Wang and Xuebing Yan
Assessment of Effects of Storage Time on Fermentation Profile, Chemical Composition, Bacterial Community Structure, Co-Occurrence Network, and Pathogenic Risk in Corn Stover Silage
Reprinted from: *Fermentation* **2025**, *11*, 425, https://doi.org/10.3390/fermentation11080425 . . . **56**

Zhaolong Li, Song Peng, Mengshi Zhao, Xiaodong Zhuang, Huini Wu, Tiecheng Sun and Fengqiang Lin
Effects of *Lactobacillus plantarum*-81-Fermented Feed on Growth and Intestinal Health of Muscovy Ducks
Reprinted from: *Fermentation* **2025**, *11*, 311, https://doi.org/10.3390/fermentation11060311 . . . **69**

Phan Vu Hai, Le Xuan Anh and Nguyen Xuan Hoa
Fermented Chive (*Allium schoenoprasum*) with *Lactobacillus plantarum*: A Potential Antibiotic Alternative Feed Additive for Broilers Challenged with *Escherichia coli*
Reprinted from: *Fermentation* **2025**, *11*, 277, https://doi.org/10.3390/fermentation11050277 . . . **89**

Peiling Wei, Mingxuan Guan, Xuhui Liang, Kaixin Yuan, Ning Chen, Yuxin Yang and Ping Gong
Growth Performance and Rumen Microbiota of Sheep Respond to Cotton Straw Fermented with Compound Probiotics
Reprinted from: *Fermentation* **2025**, *11*, 244, https://doi.org/10.3390/fermentation11050244 . . . **104**

Haozhan Qu, Gengxiu Zan, Haoyan Li, Xiaofan Wang, Jiayi Zhou, Xiuqi Wang and Huichao Yan
Fermented Feed Promotes Gut Development by Enhancing Intestinal Stem Cell Expansion via Activation of the Wnt/β-Catenin Signaling Pathway
Reprinted from: *Fermentation* **2025**, *11*, 52, https://doi.org/10.3390/fermentation11020052 . . . . **120**

Jianliang Liu, Mingjian Liu, Panjie Sheng, Chaoran Song, Weiqin Ma, Baochao Bai, et al.
Biotechnological Effects of *Lactobacillus plantarum*, Cellulase, and Xylanase on Nutritional Quality and Microbial Community Structure of Corn Stover Silage
Reprinted from: *Fermentation* **2025**, *11*, 14, https://doi.org/10.3390/fermentation11010014 . . . . **134**

Preface

Fermentation technology has stood as a cornerstone of animal nutrition for centuries, evolving from a traditional feed preservation practice to a sophisticated biotechnological tool that shapes the sustainability and efficiency of modern animal husbandry worldwide. Its capacity to transform perishable forages, under-utilized agro-industrial by-products, and low-value feed resources into palatable, nutritionally enhanced, and safe feedstuffs addresses two of the most pressing challenges in global animal agriculture: feed scarcity and the drive toward antibiotic-free, eco-friendly production systems. However, despite its longstanding use and rapid technological advancement, critical gaps persist in our understanding of the complex interplay between fermentation parameters, microbial communities, and the biological impacts of fermented feeds on animal gastrointestinal health, ruminal function, nutrient utilization, and overall productivity. These knowledge gaps limit the full optimization of fermentation technology and its translation into evidence-based practices for the animal feed industry—an unmet need that motivated the creation of this second edition of the Special Issue *Application of Fermentation Technology in Animal Nutrition*, in the Industrial Fermentation Section of *Fermentation*.

As Guest Editors of this Special Issue, we sought to curate a collection of rigorous, original research that advances both the fundamental science and practical application of fermentation in animal nutrition by building on the insights of the first edition and expanding into emerging areas of inquiry. Our aim was to bring together international researchers and practitioners to share cutting-edge findings that address key unanswered questions: How can fermentation conditions—including harvesting time, storage duration, and microbial inoculant and enzyme combinations—be optimized to enhance silage quality, aerobic stability, and nutritional value? What biotechnological strategies can unlock the potential of agricultural by-products as high-quality feed resources? What are the molecular and microbial mechanisms by which fermented feeds modulate animal gut health, immune function, and growth performance? And how can fermentation technology support the development of sustainable, antibiotic-alternative animal production systems?

This Special Issue answers these questions through ten peer-reviewed studies that span the breadth of fermentation applications in animal nutrition, from silage production optimization for corn, sorghum, and crop stover to the development of probiotic-fermented feed additives and the elucidation of fermented feed's impacts on livestock and poultry microbiota and physiological function. The research presented here offers evidence-based insights for researchers, animal nutritionists, feed industry professionals, and agricultural policymakers—our primary audience—providing actionable data on microbial inoculant efficacy, feed processing techniques, and the biological effects of fermented feeds across ruminants and monogastric animals alike.

The accompanying Reprint compiles all these seminal studies in a cohesive, accessible format, serving as a definitive reference for anyone seeking to deepen their understanding of the latest advancements in fermentation technology for animal nutrition. It is our hope that this collection not only fills critical knowledge gaps in the field but also inspires further interdisciplinary research, fosters collaboration between scientists and industry, and accelerates the adoption of sustainable fermentation-based practices in global animal husbandry. We extend our gratitude to all the authors who contributed their innovative research, the peer reviewers whose rigorous feedback ensured the scientific integrity of this issue, and the editorial team of *Fermentation* for their invaluable support throughout the process. It is our privilege to share this body of work with the community, and we look forward to the continued evolution of fermentation technology as a driving force for a more efficient, sustainable, and healthy future in animal nutrition.

Siran Wang, Lin Sun, Lei Chen, and Jie Bai
Guest Editors

fermentation

MDPI

Editorial

Application of Fermentation Technology in Animal Nutrition: 2nd Edition

Siran Wang [1,2,*], Lin Sun [3], Lei Chen [4] and Jie Bai [5]

1 Institute of Animal Science, Jiangsu Academy of Agricultural Sciences, Nanjing 210014, China
2 Key Laboratory for Crop and Animal Integrated Farming, Ministry of Agriculture and Rural Affairs, Nanjing 210014, China
3 Inner Mongolia Engineering Research Center of Development and Utilization of Microbial Resources in Silage, Inner Mongolia Academy of Agriculture and Animal Husbandry Science, Hohhot 010031, China; sunlin2013@126.com
4 College of Animal Science, Shanxi Agricultural University, Jinzhong 030801, China; cl1016zj@126.com
5 Prataculturall College, Gansu Agricultural University, Lanzhou 730070, China; baij@gsau.edu.cn
* Correspondence: wangsiran@jaas.ac.cn

Fermentation technology has long been applied in animal nutrition worldwide, with its application primarily focused on animal feed. A core objective of this technology is the long-term preservation of feedstuffs: harvested forages, perishable agro-industrial or food processing by-products, and total mixed rations containing these ingredients are frequently ensiled to ensure year-round palatability, stability, and hygiene. Another key goal of this field is to enhance the nutritional value or safety of underutilized feed resources through fermentation; without such treatment, their use as feed may be restricted due to poor digestibility or potential toxicity. To achieve these aims, a variety of physical, chemical, and microbiological techniques have been developed [1]. This Special Issue, "Application of Fermentation Technology in Animal Nutrition: 2nd Edition", includes nine original research articles and has received over 13,104 views. Collectively, these contributions demonstrate the promising prospects of applying fermentation technology in animal nutrition research—a field that integrates tradition with innovation, aiming to push the boundaries of fermentation science forward.

With the advancement of animal husbandry, the demand for quality feed has increased significantly, whereas the supply of natural forage resources remains scarce. Corn stover, featuring high yield and abundant nutritional components, has emerged as a crucial roughage source. Applying silage technology to corn stover not only enhances its palatability and nutritional value but also exerts positive impacts on livestock production performance. Liu et al. [2] assessed the effects of different additives and their dosage ratios on the nutrient composition, fermentation quality, and microbial community structure of corn stover silage. Their findings indicated that the synergistic application of cellulase, *Lactobacillus plantarum*, and xylanase substantially improved the sensory quality of corn stover silage, balanced the microbial community during the maturation period, and optimized the silage's nutritional profile. This work demonstrated that adding 20 U/g xylanase to corn stover silage significantly enhances silage quality, maturation efficiency, and microbial community stability. Therefore, the incorporation of xylanase is recommended to optimize the quality of corn stover silage.

Fermented feed is widely used in animal production owing to its enhanced palatability and high utilization efficiency. To explore the beneficial effects of feed fermentation on animal health, Qu et al. [3] conducted studies using piglet models and found that supplementation with fermented mixed feed (FMF) activates the Wnt/β—catenin signaling pathway, accelerates the expansion of intestinal stem cells (ISCs), and thereby promotes intestinal development. This finding not only provides valuable insights into the potential applications of FMF but also lays a theoretical foundation for its use as a nutritional supplement to support healthy digestive function in pigs.

To develop cotton straw into a viable feed resource via biological fermentation, Wei et al. [4] conducted a study to valorize this agricultural by-product as a novel feed and assess its feeding efficacy in sheep. The results revealed that diets containing a low proportion of fermented cotton straw (14.50% of the total feed) achieved an optimal short-term fattening effect in sheep over a 50-day trial period. Furthermore, this dietary regimen improved the sheep's utilization of fat and protein, while enhancing their antioxidant capacity. Fermented cotton straw boosted sheep production performance by increasing the abundance of nutrient-metabolizing bacteria—such as Prevotella, which is involved in the utilization of protein, carbohydrates, and other nutrients. Overall, the biological fermentation of cotton straw enables the efficient conversion of agricultural waste into animal feed, thereby facilitating sustainable agricultural development.

Escherichia coli (*E. coli*) is a prevalent bacterial pathogen in global poultry production, triggering severe health problems and declining productivity. Additionally, the overuse of antibiotics has fueled the emergence of antimicrobial resistance, which poses a grave threat to both animal and human health. To address this issue, Hai et al. [5] assessed the efficacy of fermented chive (*Allium schoenoprasum*) with *Lactobacillus plantarum* 1582 (FC) as an antibiotic alternative for controlling *E. coli* infections in broiler chickens. Their findings indicate that FC supplementation—especially at a 3% dosage—serves as a promising natural substitute for antibiotics in managing *E. coli* infections in broiler production.

Feed fermented by different strains of *Lactobacillus plantarum* (LP) generates unique bioactive substances. Li et al. [6] systematically compared the growth performance, gut microbiota regulation, and immune response indices of Muscovy ducks fed LP81-fermented feed (LP81-FF) against those on conventional diets. Their findings reveal that LP81-FF enhances Muscovy duck production via multiple mechanisms: improving nutrient bioavailability via a feed pretreatment, regulating microbiota-driven metabolic remodeling, and boosting systemic antioxidant capacity and immune function. These results confirm microbial fermentation as a feasible strategy for optimizing sustainable poultry production.

In developing tropical countries, local native grasses and agricultural by-products are the main roughage sources for ruminants. To enhance the efficient utilization of these by-products and mitigate the bottleneck of animal feed shortages during dry seasons, Du et al. [7] used corn stover (CS; *Zea mays* L.) as a raw material to systematically explore dynamic changes in the silage fermentation quality, bacterial community structure, and pathogenic risks across different fermentation durations. Their results demonstrated that CS contains suitable carbon and nitrogen sources to support natural silage fermentation: *Serratia marcescens* was the dominant microbial species in fresh CS, which gradually shifted to *Lactiplantibacillus plantarum* as fermentation progressed. Additionally, the optimal period for coordinating microbial metabolism and nutrient transformation was identified as between days 15 and 30. This study lays a theoretical foundation for optimizing the research framework of agricultural by-product bioconversion.

Whole-plant corn silage is a pivotal feedstuff in global ruminant production, with its nutrient composition tightly linked to harvest timing. Since starch serves as the main energy source in silage-based diets, exploring variations in starch degradation rates lays a theoretical foundation for optimizing the efficient use of whole-plant corn and its silage in ruminant farming. Accordingly, Zhang et al. [8] employed wet chemistry methods and

 https://doi.org/10.3390/fermentation12010058

an in vitro simulated rumen fermentation system to examine the starch content, gliadin content, amylopectin content, disappearance rate, and rumen dynamic degradation parameters of whole-plant corn silage harvested at different times. Their results revealed that silage fermentation enhances the amylopectin content in whole-plant corn starch, elevates the proportion of rapidly degradable starch, reduces gliadin levels, and increases both the rumen disappearance rate and starch disappearance rate of the silage. Additionally, the deposition of starch and gliadin in corn starches of varying structures varied with harvest timing, which in turn induced corresponding changes in the rumen degradation of whole-plant corn silage starch.

Biomass sorghum is distinguished by its high dry matter yield and ratooning capacity, supporting multiple harvests and silage production. Silva et al. [9] assessed the productive potential and fermentation quality of silages derived from ratoon biomass sorghum hybrids. Their findings revealed that these hybrids maintained high dry matter yield potential even under low-rainfall conditions, with all producing silages of satisfactory fermentation quality. The only limitation was a reduced aerobic stability, which can be alleviated via heterofermentative inoculants. Regarding chemical compositions, CMSXS5039 was the most suitable hybrid for high-energy diets, attributed to its superior starch and non-fiber carbohydrate contents. In contrast, BRS 716 and CMSXS5044 were more appropriate for high-protein diets due to their elevated crude protein levels.

Corn silage is a primary forage source for dairy cattle and a key component of intensive beef cattle production systems. Nevertheless, it is susceptible to aerobic deterioration during feedout. Microbial chitinases hold promise for the biological control of phytopathogenic fungi, as they degrade the chitin layer in fungal cell walls, offering a potential eco-friendly approach to curb such spoilage microorganisms. Thus, Niu et al. [10] assessed the impacts of a heterofermentative lactic acid bacterial inoculant—comprising *Lentilactobacillus hilgardii* and *Lentilactobacillus buchneri*—either alone or combined with chitinases, on the ensiling process, aerobic stability, and in vitro ruminal fermentation of whole-crop corn silages. Their findings indicated that inoculation with the *L. hilgardii* and *L. buchneri* mixture, both alone and in combination with chitinase, improved the aerobic stability and enhanced in vitro neutral detergent fiber digestibility of corn silages. Chitinase alone, however, exerted no such effects.

Collectively, this Special Issue features nine original studies that focus on the application of fermentation technology in animal nutrition, covering research areas such as the silage of agricultural by-products, fermented mixed feed, the isolation and application of lactic acid bacteria strains, the effect of fermented feed on animal performance, and other related topics. These findings are expected to expand feed resources, optimize fermentation quality, and enhance animal performance. While the challenges ahead are substantial, the accompanying opportunities are equally remarkable. By addressing these challenges and pursuing the research directions outlined herein, fermentation technology can be maintained as a critical component in animal nutrition, thereby meeting the evolving demands of the animal husbandry industry.

Future Challenges in the Application of Fermentation Technology in Animal Nutrition

1. Difficulties in strain and microecological regulation: Screening high-quality functional strains (e.g., high-efficiency enzyme-producing and stress-tolerant lactic acid bacteria) is time-consuming and costly. Unclear interaction mechanisms of multi-strain synergistic fermentation often cause antagonism or unstable functions, while strain inactivation in complex breeding environments impairs effect consistency.

https://doi.org/10.3390/fermentation12010058

2. Bottlenecks in process standardization and scaling-up: The precise control of solid-state fermentation parameters (temperature, humidity, pH) is challenging, leading to batch-to-batch quality variations. Liquid-state fermentation suffers from high energy consumption, excessive wastewater, and inconvenient transportation. High-temperature pelleting inactivates probiotics and active enzymes, requiring low-temperature processes or post-coating technologies that increase costs.
3. Intense safety risk prevention pressure: Fermentation is prone to pathogenic bacteria contamination (e.g., *Salmonella*) and biogenic amine or mycotoxin production. Raw material batch variations and secondary fermentation may induce nutrient loss (e.g., lysine, methionine) or reduced palatability, necessitating a full-chain detection system.
4. Inadequate regulatory and market adaptation: Unified standards for fermented feed strain evaluation, viable bacteria count detection, and safety limits are lacking. Regional regulatory differences (e.g., EU, China) in genetically modified strains and fermentation by-products raise compliance costs, and the low acceptance of fermentation technology among small- and medium-sized farms hinders its promotion.
5. Balancing effect stability and cost: Breeding conditions (stress, raw material fluctuations) and animal species/stage differences affect fermented feed efficacy. High costs from high-quality strains, specialized equipment, and long fermentation cycles are unaffordable for small- and medium-sized farms, restricting large-scale applications.

Funding: This study was financially supported by the National Natural Science Foundation of China (32301500).

Acknowledgments: We appreciate the support we received from the Special Issue "Application of Fermentation Technology in Animal Nutrition: 2nd Edition" in *Fermentation-Basel* and would like to thank authors, whose valuable work contributed to the success of this edition.

Conflicts of Interest: The authors declare no conflicts of interest.

List of Contributions

1. Wang, S. Application of Fermentation Technology in Animal Nutrition. *Fermentation* **2024**, *10*, 596.
2. Liu, J.; Liu, M.; Sheng, P.; Song, C.; Ma, W.; Bai, B.; Zhao, J.; Du, S.; Ge, G.; Wang, Z.; et al. Biotechnological Effects of Lactobacillus plantarum, Cellulase, and Xylanase on Nutritional Quality and Microbial Community Structure of Corn Stover Silage. *Fermentation* **2025**, *11*, 14.
3. Qu, H.; Zan, G.; Li, H.; Wang, X.; Zhou, J.; Wang, X.; Yan, H. Fermented Feed Promotes Gut Development by Enhancing Intestinal Stem Cell Expansion via Activation of the Wnt/β-Catenin Signaling Pathway. *Fermentation* **2025**, *11*, 52.
4. Wei, P.; Guan, M.; Liang, X.; Yuan, K.; Chen, N.; Yang, Y.; Gong, P. Growth Performance and Rumen Microbiota of Sheep Respond to Cotton Straw Fermented with Compound Probiotics. *Fermentation* **2025**, *11*, 244.
5. Hai, P.V.; Anh, L.X.; Hoa, N.X. Fermented Chive (*Allium schoenoprasum*) with Lactobacillus plantarum: A Potential Antibiotic Alternative Feed Additive for Broilers Challenged with Escherichia coli. *Fermentation* **2025**, *11*, 277.
6. Li, Z.; Pen, S.; Zhao, M.; Zhuang, X.; Wu, H.; Sun, T.; Lin, F. Effects of Lactobacillus plantarum-81-Fermented Feed on Growth and Intestinal Health of Muscovy Ducks. *Fermentation* **2025**, *11*, 311.
7. Du, Z.; Meng, Y.; Chen, Y.; Cui, S.; Wang, S.; Yan, X. Assessment of Effects of Storage Time on Fermentation Profile, Chemical Composition, Bacterial Community Structure, Co-Occurrence Network, and Pathogenic Risk in Corn Stover Silage. *Fermentation* **2025**, *11*, 425.

https://doi.org/10.3390/fermentation12010058

8. Zhang, L.; Liu, S.; Wang, X.; Wang, H.; Li, S.; Zhen, Y.; Zhang, X. Effect of Harvesting Time on Starch Degradation in Rumen of Whole-Plant Corn and Its Silage. *Fermentation* **2025**, *11*, 522.
9. Silva, Y.A.d.; Orrico Junior, M.A.P.; Retore, M.; Ceccon, G.; Amaral, I.P.d.O.; Orrico, A.C.A.; Muglia, G.R.P.; Fernandes, T. Productivity, Fermentation Parameters, and Chemical Composition of Silages from Biomass Sorghum Hybrids in Ratoon Crop. *Fermentation* **2025**, *11*, 540.
10. Niu, H.; Nair, J.; Yang, H.-E.; McAllister, T.A.; Chevaux, E.; Wang, Y. The Effects of a Heterofermentative Lactic Acid Bacterial Inoculant Containing *Lentilactobacillus hilgardii* and *Lentilactobacillus buchneri* with or Without Chitinases on the Ensiling, Aerobic Stability, and In Vitro Ruminal Fermentation of Whole-Crop Corn Silages. *Fermentation* **2026**, *12*, 29.

 https://doi.org/10.3390/fermentation12010058

fermentation

MDPI

Article

The Effects of a Heterofermentative Lactic Acid Bacterial Inoculant Containing *Lentilactobacillus hilgardii* and *Lentilactobacillus buchneri* with or Without Chitinases on the Ensiling, Aerobic Stability, and In Vitro Ruminal Fermentation of Whole-Crop Corn Silages

Huaxin Niu [1,†], Jayakrishnan Nair [2,3,†], Hee-Eun Yang [3], Tim A. McAllister [3], Eric Chevaux [4] and Yuxi Wang [3,*]

[1] College of Animal Science and Technology, Inner Mongolia Minzu University, Tongliao 028000, China; huaxin.niu@hotmail.com
[2] School of Agricultural Sciences, Southern Illinois University, Carbondale, IL 62901, USA; jay.nair@siu.edu
[3] Lethbridge Research and Development Centre, Agriculture and Agri-Food Canada, Lethbridge, AB T1J 4B1, Canada; heeeun.yang@agr.gc.ca (H.-E.Y.); tim.mcallister@agr.gc.ca (T.A.M.)
[4] Lallemand SAS, 31702 Blagnac, France; echevaux@lallemand.com
[*] Correspondence: yuxi.wang@agr.gc.ca; Tel.: +1-(403)-317-3498; Fax: +1-(403)-382-3156
[†] These authors contributed equally to this work.

Abstract

The objectives of this study were to evaluate the effects of a heterofermentative lactic acid bacterial (LAB) inoculant containing *Lentilactobacillus hilgardii* and *Lentilactobacillus buchneri* with or without chitinases on ensiling, aerobic stability (AS), and in vitro ruminal fermentation of whole-crop corn silages. Uninoculated corn silage (Control, C), corn silage treated with chitinase alone (E), with a mixture of *L. hilgardii* and *L. buchneri* (I), or with a combination of E and I (EI) were ensiled in minisilos for 90 d, and the terminal silages were assessed for AS and in vitro ruminal fermentation. The experiment used a completely randomized design, with the treatment structure allowing analysis of the effects of E, I, and the E × I interaction. I and EI had higher ($p < 0.01$) concentrations of acetate and total volatile fatty acids, higher ($p < 0.01$) total bacterial and LAB counts, but lower yeast population ($p < 0.01$) than C and E terminal silages. Aerobic stability was greater ($p < 0.001$) for I and EI (>336 h) than for E (156 h) and C (152 h) silages. LAB reduced ($p < 0.001$) while chitinase increased ($p < 0.05$) total in vitro gas production. Disappearance of neutral detergent fiber was lower ($p < 0.05$) after 6 and 12 h but higher ($p < 0.001$) after 24 (74.7 and 76.7% vs. 72.7 and 72.3%) and 48 h (84.9 and 86.2% vs. 82.9 and 80.2%) of incubation for I and EI than for C and E, respectively. Inoculation of a mixture of *L. hilgardii* and *L. buchneri* alone (I) and in combination with chitinase (EI), but not chitinase (E) alone, enhanced AS and improved the in vitro NDF digestibility of corn silages.

Keywords: heterofermentative LAB; silage nutritive value; in vitro digestibility; silage fermentation

1. Introduction

Corn silage is a major forage source for dairy cattle and an important component of intensive beef cattle operations in North America [1,2]. However, corn silage is prone to aerobic deterioration during feedout due to the activity of bacteria such as *Acetobacter* as well as yeast and mold, leading to loss of dry matter (DM) and forage nutritive value [2]. Treatment with lactic acid bacterial (LAB) inoculants is an established strategy to promote ensiling, improve silage quality, and reduce dry matter (DM) losses during storage and feedout [3–5]. Heterofermentative LAB such as *Lentilactobacillus buchneri* have been shown to be particularly effective in improving the aerobic stability of silages during feedout [4,6], owing to their ability to produce acetic acid (AC) and 1,2-propanediol, which inhibit yeasts and molds [6–11]. However, *L. buchneri* produces AC after ~60 d of ensiling [4,12,13]. In contrast, *Lentilactobacillus hilgardii*, which is also a heterofermentative LAB [4,14–17], produces AC as soon as 14 d after ensiling [6,18]. Therefore, a combination of *L. buchneri* and *L. hilgardii* could shorten the ensiling period required for enhanced aerobic stability (AS) and enable faster post-ensiling feedout for beef and dairy producers [6,17,19].

Microbial chitinases are promising candidates for biological control of phytopathogenic fungi as they degrade the chitin layer within fungal cell walls [20–22], making them a potential eco-friendly strategy to control these spoilage microorganisms [23–25]. Forages can harbor considerable populations of phytopathogenic fungi. However, studies evaluating the effects of chitinases in inhibiting fungi in conserved forages have been limited. Our previous studies have shown that application of a combination of *Pediococcus pentosaceus* or *Pichia anomala* along with chitinase during baling of high-moisture alfalfa hay reduced the internal temperature of round bales during storage and improved neutral detergent fiber (NDF) digestion in vitro [26–28]. However, the impact of a combination of heterofermentative LAB consisting of *L. buchneri* NCIMB-40788 and *L. hilgardii* CNCM I-4785 with chitinase from *Streptococcus griseus* on the ensiling, aerobic stability, nutrient digestibility, and feed value of corn silage has not been evaluated. It was hypothesized that chitinase, with or without LAB inoculants, would not negatively impact microbial fermentation during ensiling or aerobic exposure but would improve in vitro NDF degradability. The objectives of this study were to evaluate the effects of a heterofermentative LAB inoculant consisting of *L. hilgardii* and *L. buchneri* with or without chitinases on the ensiling, AS, and in vitro ruminal fermentation of whole-crop corn silages.

2. Materials and Methods

2.1. Forage, Treatments, and Ensiling

Corn (*Zea mays*) was planted under irrigation at the Lethbridge Research and Development Center, AB, Canada, in May 2018. The seeds were planted at a depth of 5–6 cm with a seeding density of 89,500 seeds ha^{-1}. Whole-crop corn was harvested in October 2018 at two-thirds milk line (33.9 $\pm$ 0.87% DM), as described by Nair et al. [6]. The harvested forage was chopped to ca. 9.5 mm theoretical length with a Claas Jaguar 940 forage harvester (Claas KGaA mbH, Harsewinkel, Germany) equipped with a kernel processor, with rollers adjusted to 1.0 mm clearance.

Treatments were arranged in a completely randomized design, allowing for the evaluation of the effects of E, I, and the E $\times$ I interaction. Treatments included uninoculated control corn silage (C), and corn silages treated with 0.5 mg/kg fresh forage chitinase (E) from *Streptococcus griseus* ($\geq$200 chitinase units/g, Sigma-Aldrich, St. Louis, MO, USA) alone (E), LAB inoculant (Lallemand Specialties Inc., Milwaukee, WI, USA) containing 1.5 $\times$ 10^8 [29] colony forming units (cfus)/kg fresh forage *L. hilgardii* CNCM I-4785 and 1.5 $\times$ 10^8 cfu/kg fresh forage *L. buchneri* NCIMB-40788 (I), or a combination of E and I (EI). One unit of chitinase was defined as the amount of enzyme that liberated

 https://doi.org/10.3390/fermentation12010029

1.0 mg of N-acetyl-D-glucosamine from chitin per h at a pH of 6.0 at 25 °C in a 2 h assay [30].

2.2. *Preparation of Whole-Crop Corn Silage in Minisilos*

Minisilos were prepared as described by Nair et al. [31]. Briefly, harvested corn forage was divided into twenty 25 kg lots, with five lots per treatment, and placed on separate clean plastic sheets. Corn forage was then treated with chitinase, LAB inoculant, or a combination by spraying the respective inoculants diluted in distilled water at 25 mL/lot to provide the required application rates, as specified above. The C silage was sprayed with 25 mL of double-distilled water. Each lot was thoroughly hand-mixed to ensure uniform distribution of additives. In order to prevent cross-contamination, separate containers and sprayers were used, and gloves were changed between treatments. Mixing and application were repeated for each of the replicate lots, resulting in a total of five mixes per treatment.

After mixing, exactly 2.0 kg of forage from each replicated lot was weighed and packed into each of three minisilos (PVC pipes having 10.4 cm in width × 35.6 cm in length and rubber end caps), with one for each ensiling duration (7, 21, and 90 d), using a hydraulic press equipped with a pressure gauge to achieve the recommended packing density of ~240 kg DM/m^3 for proper ensiling. After packing, minisilos were sealed with the rubber end caps having the air venting tubes closed with binder clips and secured with 4-inch metal hose clamps. A total of 60 minisilos were prepared, with 5 replicate minisilos per treatment for each ensiling duration. Each labeled minisilo was weighed immediately after filling and sealing, with the weights recorded along with the weight of empty minisilos, and was packed in Rubbermaid containers and stored indoors at ambient temperature (20 °C) throughout the ensiling period. The minisilos were vented every day initially, followed by periodic venting until there was no significant gas production (~2 weeks of storage). Five replicate samples of fresh forage were collected per treatment prior to packing of minisilos on the day of ensiling (d 0) for chemical and microbial analysis. On each sampling day, prior to opening, the minisilos were weighed, and the weights recorded were used to calculate DM losses. After opening the minisilos, the content of each minisilo was thoroughly mixed and subsampled for chemical and microbial analyses.

2.3. *Aerobic Stability*

Silage samples from minisilos opened at the end of 90 d of ensiling were used to assess AS. About 1.2 kg each of the terminal silage subsamples from the five replicate minisilos per treatment were mixed thoroughly and packed into 4 L insulated containers (n = 3; 13.5 cm in diameter × 30.9 cm in height). The containers were covered with two layers of cheesecloth and stored at ambient temperature (20 °C) for 14 d. Two Dallas Thermochron iButton sensors (Embedded Data Systems, Lawrenceburg, KY, USA) were placed in the silage within each container at ~9.0 cm and ~18.0 cm from the base of the container to measure temperature every 15 min throughout the 14 d of storage. Two sensors were also used to measure the ambient room temperature. After 3, 7, and 14 d of aerobic exposure (AE), the contents of each container were sub-sampled to evaluate the pH, chemical, and microbial composition. Aerobic stability was determined as the duration (hours) before the temperature of silage subjected to aerobic exposure exceeded the ambient temperature by 2 °C [32].

2.4. *In Vitro Experiment*

Three ruminally cannulated Angus heifers (550 ± 50 kg body weight) were used as rumen fluid donors for this experiment. The heifers were fed a total mixed ration

https://doi.org/10.3390/fermentation12010029

consisting of 65.0% corn silage, 20.0% barley grain, 10.0% canola meal, and 5.0% of a vitamin-mineral supplement (% DM basis). The diet was formulated to meet or exceed the National Academies of Sciences, Engineering, and Medicine (NASEM) [33] nutrient requirements for beef heifers. All heifers were fed at 08:00 h and provided ad libitum access to feed and water and cared for in accordance with the standards of the Canadian Council on Animal Care (LRC protocol ACC1830) [34]. Rumen fluid was collected 2 h after the morning feeding from five locations within the rumen, strained through four layers of cheesecloth, combined in equal volumes among three cattle, and immediately transported in an anaerobic and pre-warmed container to the laboratory. Rumen fluid was then combined (1:2, v/v) with pre-warmed (39 °C) mineral buffer [35] to generate the inoculum.

2.5. In Vitro Incubations and Measurements

Terminal silage samples (d 90 of ensiling) from all four treatments were dried in a forced-air oven at 55 °C for 48 h and ground through a 1 mm screen using a Wiley Mill (Model 4, Arthur H. Thomas Co., Philadelphia, PA, USA). Ground samples (0.5 g each) were weighed in triplicate in acetone-rinsed Ankom F57 filter bags (25 µm porosity; Ankom Technology Crop., Macedon, NY, USA), heat-sealed, and placed in 125 mL capacity serum vials. Pre-warmed inoculum was dispensed under a stream of O_2-free CO_2 into pre-warmed serum vials (60 mL/vial), which were immediately sealed and placed on a rotary shaker platform (140 rev/min) in an incubator at 39 °C. Vials containing only rumen inoculum were incubated as blanks. The 0 h samples were immediately placed into an ice-water bath after preparation. The experiment was repeated twice, one week apart. Gas production (GP) from each vial was measured after 3, 6, 12, 24, and 48 h of incubation, by inserting a 26-gauge (0.4 mm) needle attached to a three-way stopcock and a pressure transducer (model PX4200-015GI, Omega Engineering Inc., Laval, QC, Canada) connected to a visual display device (Data Track, Christchurch, UK). After 48 h, fermentation was inhibited by removing the serum vials from the incubator and immersing them in ice-water. The vials were opened to remove the filter bags, the pH of the fermentation liquid was then measured with a Symphony pH meter (SB70P, VWR, Mississauga, ON, Canada), and two 1 mL samples were preserved with 200 µL of 25% (w/v) metaphosphoric acid or 200 µL of 1% (v/v) aqueous 18.4 M sulfuric acid and stored at −20 °C for analysis of volatile fatty acids (VFAs) and ammonia nitrogen (NH_3-N), respectively. After removal from the vials, filter bags were rinsed with cold water until the water was clear and dried in a forced-air oven at 55 °C for 48 h. Bags were weighed to estimate DM disappearance (DMD) based on the amount of DM lost from the bags after incubation. Residues remaining in the filter bags were further analyzed individually for neutral detergent fiber (NDF), as described below, which was used to determine in vitro NDF disappearance (NDFD).

2.6. Microbial Analyses of Corn Silage

Microbial analyses of silage were conducted as described previously by Nair et al. [6] and Addah et al. [36]. Briefly, fresh corn forage samples collected on the day of ensiling, silage samples from minisilos at each opening day, and aerobically exposed silage samples collected during each sampling day during the 14 d of aerobic exposure (10 g each) were blended and homogenized with 90 mL of sterile 70 mM potassium phosphate buffer (pH = 7.0) for 30 s at 260 rpm in a Stomacher™ Model 400 Circulator Lab Blender (Seward Medical Limited, London, UK). The suspension was serially diluted (10^{-1} to 10^{-7}) using the same buffer and 100 µL aliquots of three selected dilutions expected to result in enumerable microbial colonies were spread in triplicate onto de Man, Rogosa, and Sharpe agar (MRS;

 https://doi.org/10.3390/fermentation12010029

Dalynn Biologicals, Calgary, AB, Canada) for LAB, onto nutrient agar (NA; Dalynn Biologicals, Calgary, AB, Canada) for total bacteria (TB) and onto Sabouraud's dextrose agar (SDA; Dalynn Biologicals, Calgary, AB, Canada) for yeast and molds. Both MRS and NA plates were pre-prepared with 200 μg/mL of cycloheximide (Dalynn Biologicals, Calgary, AB, Canada), while SDA plates contained 100 μg/mL each of tetracycline and chloramphenicol. Both MRS and NA plates were incubated in a Fisherbrand™ Isotemp™ Microbiological Incubator (Thermo Fisher Scientific, Pittsburgh, PA, USA) at 37 °C for 24–48 h, while SDA plates were incubated at ambient temperature for 72 h. Plates containing a minimum of 30 and a maximum of 300 colonies were enumerated for microbial evaluation.

2.7. Chemical Analysis

Chemical analyses of fermentation products were conducted as described previously by Nair et al. [6] and Addah et al. [36]. Briefly, samples (15 g each) collected as described previously for microbial analysis were mixed with 135 mL of deionized water and blended for 30 s at full speed in a commercial blender (Waring Commercial, Torrington, CT, USA). The suspension was filtered through two layers of cheesecloth, and the pH of the collected fluid was measured in duplicate using a pH meter as described in the in vitro assessment. One portion (7.5 mL) of the filtrate was boiled immediately for 10 min to stop fermentation and stored at −20 °C for the analysis of water-soluble carbohydrates (WSCs) following the Nelson–Somogyi method [37] using a Dynatech MRX micro-plate reader (Dynatech Laboratories Inc., Chantilli, VA, USA). The second portion was stored on ice for the analysis of fermentation products. Samples were centrifuged at 10,000× *g* for 15 min at 4 °C in a Sorvall Legend Mach 1.6R centrifuge (Thermo electron Corporation, Gormley, ON, USA). The collected supernatant was analyzed, as described by Addah et al. [36], for analysis of VFA using a Hewlett Packard model 5890A Series Plus II gas–liquid chromatograph (Hewlett Packard Co., Palo Alto, CA, USA), lactate (LA), and ammonia nitrogen (NH_3-N) following the procedures of Kudo et al. [38] and Broderick and Kang [39], respectively.

Proximate analysis of silage samples was conducted following the Association of Official Analytical Chemists (AOAC) [40] procedures. Briefly, samples were dried in a forced air oven at 55 °C for 48 h, followed by grinding to pass through a 1 mm screen of a Wiley Mill (Model 4, Arthur H. Thomas Co., Philadelphia, PA, USA). Samples were analyzed for DM (Method 930.15), CP (Method 990.03), NDF, as per Van Soest et al. [41], with the addition of sodium sulfite and heat-stable α-amylase, ADF (Method 973.18), ash (Method 942.05), starch, following Herrera-Saldana et al. [42], using Megazyme® (Megazyme Int. Ltd., Wicklow, Ireland)—as described by Addah et al. [43]—and WSC, as per Snell and Snell [44], as described by Hristov et al. [45].

2.8. Calculation and Statistical Analysis

Microbial populations were log-transformed before statistical analysis and estimated as cfu/g silage DM. The *DM loss* from minisilos during ensiling was calculated as:

$$DM\ loss\ (\%) = 100 - [(WO \times DMO)/(WC \times DMC)] \times 100$$

where *WO* = weight of forage at opening, *DMO* = DM content at opening, *WC* = weight of forage at closure, and *DMC* = DM content at closure [46]. Silage temperature parameters evaluated included the maximum temperature (T_{max}), duration (h) when the silage temperature was above the ambient temperature +2 °C during AE (threshold for spoilage), and the area above ambient temperature +2 °C. T_{max} was the highest temperature recorded during 14 d of AE. The duration (h) was measured by adding the length of time (h) the

 https://doi.org/10.3390/fermentation12010029

silage temperature was above ambient temperature +2 °C during AE. The area above ambient temperature +2 °C was calculated for d 1–3, 1–7, 8–14, and 1–14 by multiplying the duration (h) above the ambient temperature +2 °C by the temperature above the ambient temperature +2 °C during AE.

Gas production data from in vitro incubation were fitted to a nonlinear model proposed by Lopez et al. [47]:

$$G = A \times [1 - e^{-c(t-Lag)}]$$

where G (mL) is the cumulative gas production, A (mL/g DM) represents the asymptotic gas production, c (/h) is the fractional fermentation rate, Lag (h) is the initial delay in the onset of gas production, and t (h) is the time of measurement. The parameters A, c, and Lag were estimated by an iterative least squares procedure using NLIN of SAS (2012).

All data were analyzed by the analysis of variance using the mixed model procedure of SAS (version 9.3.1; SAS Inst. Inc.) [48] using the following model:

$$Yij = \mu + Ti + \varepsilon ij$$

where Yij is the observation of the dependent variable, μ is the overall mean, Ti is the fixed effect of treatment, and εij is the random error associated with the observation. The duration of ensiling, AE, and the hours of in vitro incubation were the random effects. The effects of E, I, and E × I interaction were included in the model. Minisilos (n = 5), insulated containers (n = 3), and serum vials (n = 3) served as the experimental units for ensiling, AE, and in vitro incubations, respectively. For the minisilo experiment, the effects of treatment on nutrient composition, fermentation parameters, and microbial data during ensiling or AE were assessed using a repeated measure analysis with the effects of treatment (T), days of ensiling or AE (D), and T × D included in the model. Means were separated using Tukey's test. Significant differences and trends were declared at $p < 0.05$ and $0.10 > p > 0.05$, respectively.

3. Results

3.1. Fermentation Characteristics During Ensiling and Aerobic Exposure

Chemical composition, fermentation products, and microbial population of fresh corn forage and terminal (90 d of ensiling) silages ensiled in minisilos are presented in Table 1. There was no I × E interaction for any of the nutrient composition parameters evaluated. Terminal silage DM, OM, CP, NDF, ADF, and starch did not vary ($p > 0.05$) among treatments. Compared to C and E, the application of inoculant (for I and EI) decreased ($p < 0.01$) the WSC concentration of terminal silages.

Regarding the kinetics of fermentation, there was a T × D interaction for silage pH (Figure 1A), with the pH of C being lower (3.49 vs. 3.60; $p < 0.05$) than that of EI after 90 d of ensiling, whereas the pH of C was lower ($p < 0.05$) than that of I, E, and IE on d 3 (3.48 vs. 3.56, 3.60 and 3.62, respectively), greater (p < 0,05) than that of I and EI on d 7 (3.93 vs. 3.52 and 3.60, respectively), and greater ($p < 0.05$) than that of I, E, and EI on d 14 (8.31 vs. 3.92, 7.66 and 3.73, respectively) of AE.

Concentrations of acetic acid (AC; Figure 1B) were greater for ($p < 0.01$) I and EI but not for E over 90 d of ensiling, leading to higher ($p < 0.01$) concentration of AC for I and EI than for C and E on d 21 and 90 of ensiling and throughout AE. As AC accounted for ~95% of total VFA, the effects of treatment on total VFA followed the same trend as both ensiling and AE. Concentrations of propionate and butyrate were not affected by treatments. However, LA concentrations were greater ($p < 0.05$) for I than for C on d 21 ($p < 0.01$, Figure 1C) and for C and I than for E and EI on d 90 of ensiling. Concentrations of

 https://doi.org/10.3390/fermentation12010029

lactic acid (LA) during 14 d of AE remained similar for I and EI but decreased for C and E. LA concentrations of EI were lower than C on d 3 ($p < 0.05$), but greater than C on d 7 and 14 of AE ($p < 0.01$). The LA:AC ratio ($p < 0.01$) and ethanol concentrations ($p < 0.05$) of C and E were greater than those of I and EI during ensiling and AE. Succinate concentrations were lower ($p < 0.01$) for E and EI than for I.

Table 1. Chemical composition, fermentation products, and microbial populations of fresh corn forage, corn silage treated with distilled water (C), treated with chitinase (E), inoculant (I) containing *L. buchneri* and *L. hilgardii*, or with E and I combination (EI), ensiled for 90 d in minisilos.

Item [5]	Corn Forage [1]	Silage After 90 d of Ensiling [2]					*p*-Value [4]		
		C	E	I	EI	SEM [3]	I	E	I × E
pH	5.84 ± 0.104	3.49^{b}	3.55^{ab}	3.54^{bc}	3.60^{a}	0.018	0.015	0.005	0.257
DM	33.9 ± 0.87	33.6	35.0	33.5	34.4	0.68	0.245	0.121	0.227
OM, % DM	95.5 ± 0.15	96.0	95.6	96.1	96.3	0.30	0.249	0.789	0.337
CP, % DM	8.19 ± 0.22	8.12	8.02	8.16	8.45	0.52	0.28	0.261	0.131
ADF, % DM	25.6 ± 2.00	26.7	28.3	28.1	26.1	0.98	0.536	0.783	0.097
NDF, % DM	44.2 ± 1.63	44.6	47.2	48.1	45.5	1.27	0.377	0.995	0.118
Starch, % DM	23.7 ± 2.17	22.3	22.5	20.8	23.4	1.38	0.337	0.136	0.154
DM loss, %	NA	6.77	2.87	4.49	4.41	2.16	0.862	0.368	0.388
WSC, mg/g DM	43.4 ± 6.51	30.5^{a}	28.4^{a}	6.4^{b}	7.7^{b}	1.47	<0.001	0.803	0.236
Fermentation products, mg g^{-1} DM									
Acetate (AC)	NA	16.4^{b}	15.7^{b}	31.0^{a}	29.5^{a}	1.45	<0.001	0.224	0.422
Propionate	NA	0.36	0.34	0.35	0.34	0.010	0.640	0.112	0.875
Butyrate	NA	0.32	0.30	0.39	0.31	0.027	0.214	0.103	0.335
Total VFA	NA	17.4^{b}	16.7^{b}	32.0^{a}	28.7^{a}	1.49	<0.001	0.214	0.425
Lactate (LA)	NA	63.6^{a}	49.9^{b}	66.3^{a}	44.9^{c}	1.66	0.426	<0.001	0.027
LA:AC ratio	NA	3.90^{a}	3.21^{b}	2.16^{c}	1.61^{c}	0.201	<0.001	0.015	0.761
Succinate	NA	0.95^{ab}	0.81^{b}	1.09^{a}	0.79^{b}	0.056	0.293	0.004	0.193
Ethanol	NA	0.55^{ab}	0.72^{a}	0.25^{b}	0.32^{b}	0.108	0.011	0.303	0.644
NH_3-N	NA	1.11	0.97	1.02	0.98	0.048	0.427	0.097	0.270
Microbial population (log_{10} CFU g^{-1}; fresh basis)									
Total bacteria	9.05 ± 1.307	8.10^{c}	8.18^{c}	8.86^{b}	9.08^{a}	0.066	<0.001	0.033	0.280
LAB	5.74 ± 0.293	7.93^{d}	8.23^{c}	8.87^{b}	9.12^{a}	0.074	<0.001	0.002	0.721
Yeasts	7.65 ± 0.131	5.54^{a}	5.71^{a}	ND	1.11^{b}	0.377	<0.001	0.108	0.234
Mold	6.61 ± 0.267	1.09	0.91	ND	ND	0.625	-	-	-

[1] Values for fresh forage were not included in statistical analysis ($n = 3$). [2] Treatments included C, uninoculated control silage; E, corn silage ensiled with 0.5 mg/kg fresh forage chitinase; I, corn silage inoculated (cfu/g fresh forage) with lactic acid bacterial (LAB) inoculant of 1.5×10^8 *L. buchneri* and 1.5×10^8 *L. hilgardii* for a total of 3.0×10^8 LAB; EI, corn silage inoculated with combination of E and I. [3] SEM, pooled standard error of mean ($n = 5$). [4] *p*-values are for the effect of treatments having LAB inoculant (I and EI), chitinase enzyme (E and EI), and the interaction between LAB inoculant and chitinase (I × E). [5] DM, dry matter; OM, organic matter; CP, crude protein; ADF, acid detergent fiber; NDF, neutral detergent fiber; VFA, volatile fatty acids; LAB, lactic acid bacteria; LA:AC ratio, ratio of lactic acid to acetic acid; NA, not applicable; ND, not detected; NH_3-N, ammonia nitrogen; WSC, water-soluble carbohydrate. [a,b,c,d] within a row means without a common letter differ ($p < 0.05$) among treatments within ensiling.

Total bacterial counts (log_{10} cfu/g DM) in I (9.5) and EI (9.5) were greater ($p < 0.05$) than those in C (8.59) and E (8.63) after 21 d of ensiling. Similarly, TB counts in I (8.74) and EI (9.08) were greater ($p < 0.05$) than those in C (8.10) and E (8.18) after 90 d of ensiling (Figure 2A). The TB counts of EI remained greater ($p < 0.05$) than those of C and E on d 3 (9.92 vs. 8.24 and 8.25, respectively) and 7 (8.99 vs. 8.27 and 8.11, respectively)

 https://doi.org/10.3390/fermentation12010029

of AE, while the numbers of TB were lower ($p < 0.01$) for I (8.43) and EI (8.67) than for C (9.26) and E (9.41) on d 14 of AE. Similarly, LAB counts ($\log_{10}$ cfu/g DM) in I (9.44) and EI (9.35) were higher ($p < 0.05$) than those in C (8.60) and E (8.62) after 21 d of ensiling. Similarly, LAB counts in I (8.87) and EI (9.12) were higher ($p < 0.05$) than those in C (7.93) and E (8.32) after 90 d of ensiling (Figure 2B). Moreover, the LAB counts of EI remained higher ($p < 0.05$) than those of E on d 3 (9.04 vs. 8.21) and 7 (8.98 vs. 8.17) of AE, while the LAB counts of C (10.16) and E (10.19) were higher than those of I (8.54) on d 14 of AE. Addition of I and EI decreased ($p < 0.01$) yeast population, as compared to C and E silage after 90 d of ensiling (0.0 and 1.11 vs. 5.54 and 5.71, respectively) and throughout AE (Figure 2C). Mold was not detected in terminal silages and during AE across treatments (Figure 2D). Chitinase alone did not affect ($p > 0.05$) any of these microbes during ensiling and AE.

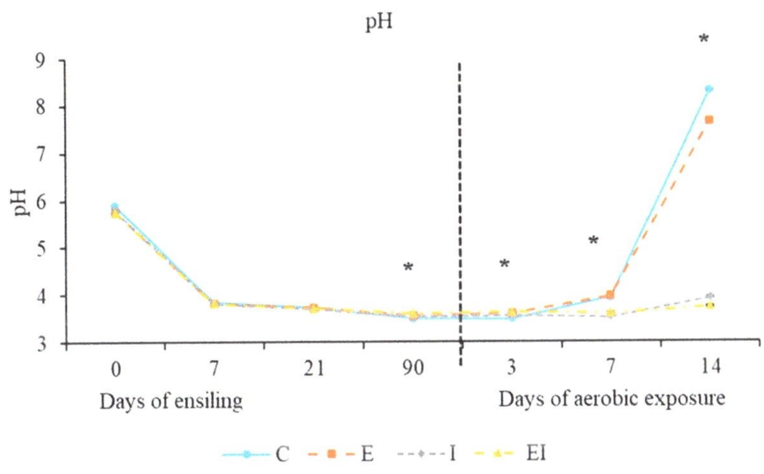

(**A**)

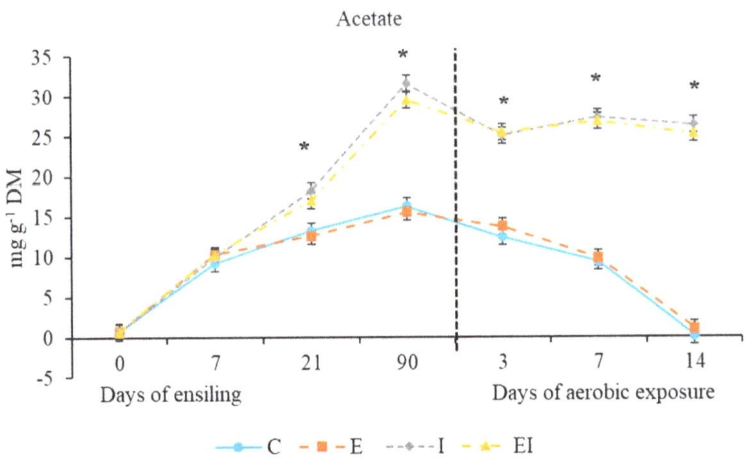

(**B**)

Figure 1. *Cont.*

https://doi.org/10.3390/fermentation12010029

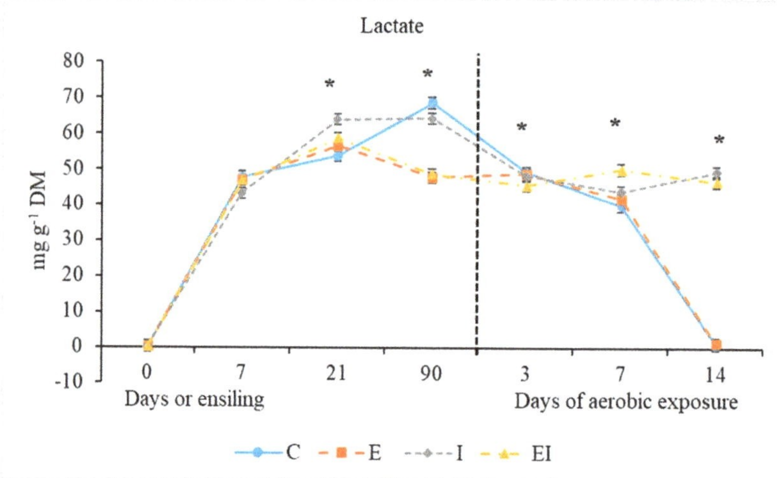

(C)

Figure 1. (**A**) Impact of addition of distilled water (C), 0.5 mg/kg fresh forage chitinase (E), lactic acid bacterial (LAB) inoculant (I) of 1.5 × 10^8 *Lactobacillus buchneri* and 1.5 × 10^8 *L. hilgardii* for a total of 3.0 × 10^8 (cfu/g fresh forage) LAB, or combination of E and I (EI) to whole-crop corn silage on pH during 90 d ensiling and 14 d aerobic exposure (AE). Area a to the left of the dotted vertical line indicates ensiling, and to the right indicates the period of AE. * Treatment difference ($p < 0.05$) during ensiling and AE. Statistical analysis was conducted independently for each of these two phases. (**B**) Impact of addition of distilled water (C), 0.5 mg/kg fresh forage chitinase (E), lactic acid bacterial (LAB) inoculant (I) of 1.5 × 10^8 *Lactobacillus buchneri* and 1.5 × 10^8 *L. hilgardii* for a total of 3.0 × 10^8 (cfu/g fresh forage) LAB, or combination of E and I (EI) to whole-crop corn silage on acetate concentrations during 90 d ensiling and 14 d aerobic exposure (AE). Area a to the left of the dotted vertical line indicates ensiling, and to the right indicates the period of AE. * Treatment difference ($p < 0.05$) during ensiling and AE. Statistical analysis was conducted independently for each of these two phases. (**C**) Impact of addition of distilled water (C), 0.5 mg/kg fresh forage chitinase (E), lactic acid bacterial (LAB) inoculant (I) of 1.5 × 10^8 *Lactobacillus buchneri* and 1.5 × 10^8 *L. hilgardii* for a total of 3.0 × 10^8 (cfu/g fresh forage) LAB, or combination of E and I (EI) to whole-crop corn silage on lactate concentrations during 90 d ensiling and 14 d aerobic exposure (AE). Area a to the left of the dotted vertical line indicates ensiling, and to the right indicates the period of AE. * Treatment difference ($p < 0.05$) during ensiling and AE. Statistical analysis was conducted independently for each of these two phases.

Aerobic stability of the terminal silages was increased ($p < 0.001$) by I, but not by E alone (Table 2). Aerobic stability of I and EI was greater ($p < 0.001$) than that of C and E during 14 d AE. The temperature of I and EI was close to ambient from d 0 to 14 of AE, while that of C and E increased after 5 d of AE (Figure 3).

Maximum temperature (T_{max}, °C) was higher ($p < 0.001$), and the time taken to reach T_{max} (h) was greater ($p < 0.001$) for C and E than for I and EI during AE (Table 2). The duration (h) that silage temperatures were above the +2 °C threshold of ambient temperature was lower ($p < 0.001$) for I and EI than for C and E when measured for d 1–3, 1–7, 8–14, and 1–14 of AE. It should be noted that the T_{max} of I and EI were close to ambient temperature. Moreover, the area [duration (h) × temperature] under the curve when the silage temperatures were above (+2 °C) ambient was greater ($p < 0.01$) for C and E than for I and EI on days 1–3, 1–7, 8–14, and 1–14 of AE. There was a significant E × I interaction ($p < 0.001$) with the area under the curve during d 1–14 and 8–14 of AE, with a greater area with the application of E alone ($p < 0.001$) but no impact with EI.

https://doi.org/10.3390/fermentation12010029

Table 2. Temperature data of corn silage treated with distilled water (C), with chitinase (E), inoculant (I) containing *L. buchneri* and *L. hilgardii*, or with E and I combination (EI), ensiled in minisilos during 14 days of aerobic exposure (AE).

	Treatments [1]					*p*-Value [3]		
Item [4]	**C**	**E**	**I**	**EI**	**SEM [2]**	**I**	**E**	**I × E**
Aerobic stability (h)	152.0 [b]	156.0 [b]	>336.0 [a]	>336.0 [a]	1.91	0.003	0.328	0.328
T_{max}, °C	33.6 [a]	34.7 [a]	18.8 [b]	18.9 [b]	0.45	<0.001	0.234	0.301
Time to reach T_{max}, h	187.3 [a]	194.7 [a]	90.7 [b]	18.7 [c]	18.86	<0.001	0.125	0.069
Duration, h								
Days 1–3	-	-	-	-	-	-	-	-
Days 1–7	16.0 [a]	12.0 [a]	0 [b]	0 [b]	1.92	<0.001	0.328	0.328
Days 8–14	167.6 [a]	167.5 [a]	0 [b]	0 [b]	0.05	<0.001	0.347	0.347
Days 1–14	183.6 [a]	179.5 [a]	0 [b]	0 [b]	1.92	<0.001	0.320	0.320
Area, duration (h) × temperature								
Days 1–3	-	-	-	-	-	-	-	-
Days 1–7	10.8 [a]	5.7 [ab]	0 [b]	0 [b]	1.99	0.003	0.236	0.236
Days 8–14	703.9 [b]	1291.4 [a]	0 [c]	0 [c]	43.08	<0.001	<0.001	<0.001
Days 1–14	714.7 [b]	1297.1 [a]	0 [c]	0 [c]	41.66	<0.001	<0.001	<0.001

[1] Treatments included C, uninoculated control silage; E, corn silage ensiled with 0.5 mg/kg fresh forage chitinase; I, corn silage inoculated (cfu/g fresh forage) with lactic acid bacterial (LAB) inoculant of 1.5×10^8 *L. buchneri* and 1.5×10^8 *L. hilgardii* for a total of 3.0×10^8 LAB; EI, corn silage inoculated with combination of E and I. [2] SEM, pooled standard error of mean ($n = 4$). [3] *p*-values are for the effect of treatments having LAB inoculant (I and EI), chitinase enzyme (E and EI), and the interaction between LAB inoculant and chitinase (I × E). [4] T_{max}, maximum temperature attained during AE; time to reach T_{max}, time (h) taken to reach maximum temperature during AE; duration, time (h) the treatment silages were above ambient temperature +2 °C during days 1–3, 1–7, 8–14, and 1–14 of AE; area, area (duration (h) × temperature) under the curve when the silage temperatures were above ambient temperature +2 °C during days 1–3, 1–7, 8–14, and 1–14 of AE. [a,b,c] within a row means without a common letter differ ($p < 0.05$) among treatments.

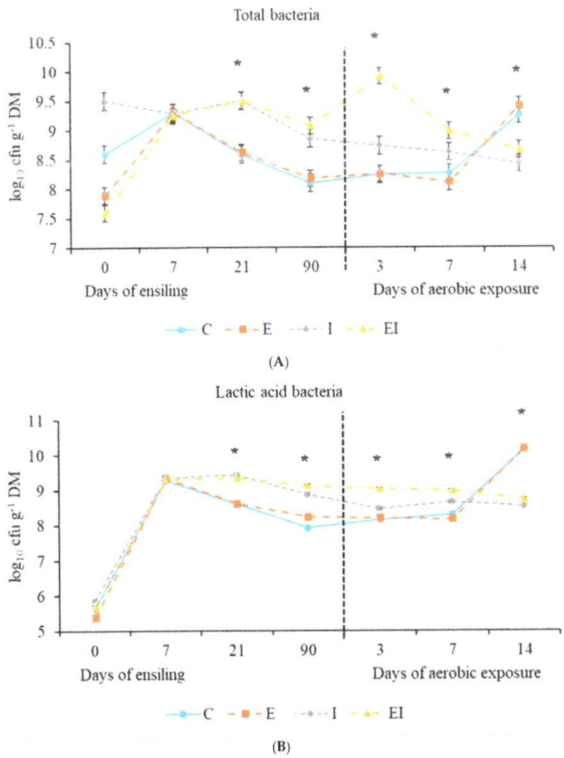

Figure 2. *Cont.*

https://doi.org/10.3390/fermentation12010029

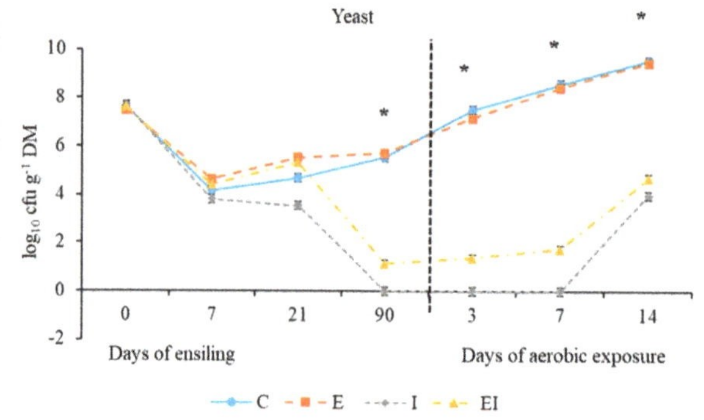

(C)

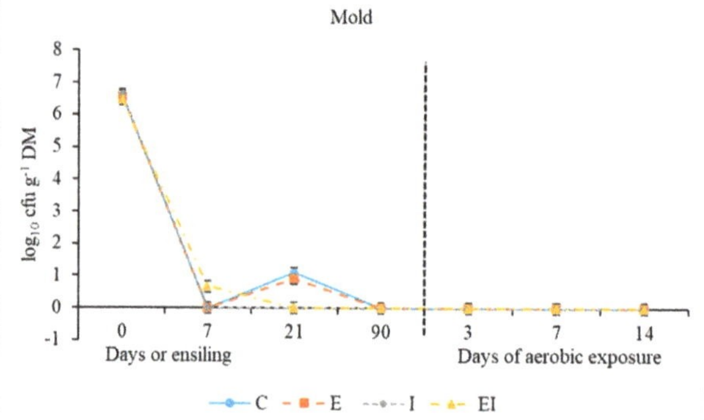

(D)

Figure 2. (**A**) Impact of addition of distilled water (C), 0.5 mg/kg fresh forage chitinase (E), lactic acid bacterial (LAB) inoculant (I) of 1.5×10^8 *Lactobacillus buchneri* and 1.5×10^8 *L. hilgardii* for a total of 3.0×10^8 (cfu/g fresh forage) LAB, or combination of E and I (EI) to whole-crop corn silage on total bacterial counts during 90 d ensiling and 14 d aerobic exposure (AE). Area a to the left of the dotted vertical line indicates ensiling, and to the right indicates the period of AE. * Treatment difference ($p < 0.05$) during ensiling and AE. Statistical analysis was conducted independently for each of these two phases. (**B**) Impact of addition of distilled water (C), 0.5 mg/kg fresh forage chitinase (E), lactic acid bacterial (LAB) inoculant (I) of 1.5×10^8 *Lactobacillus buchneri* and 1.5×10^8 *L. hilgardii* for a total of 3.0×10^8 (cfu/g fresh forage) LAB, or combination of E and I (EI) to whole-crop corn silage on lactic acid bacterial counts during 90 d ensiling and 14 d aerobic exposure (AE). Area a to the left of the dotted vertical line indicates ensiling, and to the right indicates the period of AE. * Treatment difference ($p < 0.05$) during ensiling and AE. Statistical analysis was conducted independently for each of these two phases. (**C**) Impact of addition of distilled water (C), 0.5 mg/kg fresh forage chitinase (E), lactic acid bacterial (LAB) inoculant (I) of 1.5×10^8 *Lactobacillus buchneri* and 1.5×10^8 *L. hilgardii* for a total of 3.0×10^8 (cfu/g fresh forage) LAB, or combination of E and I (EI) to whole-crop corn silage on yeast counts during 90 d ensiling and 14 d aerobic exposure (AE). Area a to the left of the dotted vertical line indicates ensiling, and to the right indicates the period of AE. * Treatment difference ($p < 0.05$) during ensiling and AE. Statistical analysis was conducted independently for each of these two phases. (**D**) Impact of addition of distilled water (C), 0.5 mg/kg fresh forage chitinase (E), lactic acid bacterial (LAB) inoculant (I) of 1.5×10^8 *Lactobacillus buchneri* and 1.5×10^8 *L. hilgardii* for a total of 3.0×10^8 (cfu/g fresh forage) LAB or combination of E, and I (EI) to whole-crop corn silage on mold counts during 90 d ensiling and 14 d aerobic exposure (AE). Area a to the left of the dotted vertical line indicates ensiling, and to the right indicates the period of AE. * Treatment difference ($p < 0.05$) during ensiling and AE. Statistical analysis was conducted independently for each of these two phases.

https://doi.org/10.3390/fermentation12010029

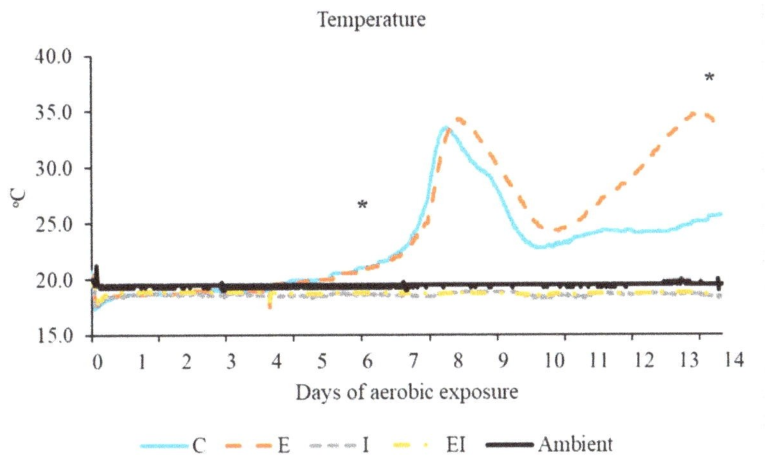

Figure 3. Impact of addition of distilled water (C), 0.5 mg/kg fresh forage chitinase (E), lactic acid bacterial (LAB) inoculant (I) of 1.5 × 10^8 *Lactobacillus buchneri* and 1.5 × 10^8 *L. hilgardii* for a total of 3.0 × 10^8 (cfu/g fresh forage) LAB, or combination of E and I (EI) to whole-crop corn silage on silage temperature during 14 d aerobic exposure. * Treatment difference ($p < 0,05$) during 14 d aerobic exposure.

3.2. In Vitro Ruminal Fermentation

The kinetic parameters of GP, NDFD, as well as VFA production during 48 h in vitro ruminal fermentation of terminal silage are presented in Table 3. No I × E interaction was observed on GP kinetic parameters. Inoculant reduced ($p < 0.001$) while E increased ($p < 0.05$) total gas production (A) during 48 h of incubation. Inoculated silages also had a lower ($p < 0.001$) rate of GP (c) and shorter ($p < 0.01$) *Lag* time compared to C silage. In contrast, E had no effect on c but increased ($p < 0.05$) *Lag* time.

Table 3. Gas production, neutral detergent fiber disappearance (NDFD), and volatile fatty acid (VFA) profile of corn silage incubated for 48 h with mixed rumen microbes.

	Treatments [1]					*p*-Value [3]		
Item	**C**	**E**	**I**	**EI**	**SEM** [2]	**I**	**E**	**E × I**
			Gas production kinetics [4]					
A (ml g^{-1} DM)	305.8 ^{a}	314.4 ^{a}	282.7 ^{b}	290.4 ^{b}	16.61	<0.001	0.011	0.896
c (h^{-1})	0.185 ^{a}	0.189 ^{a}	0.153 ^{b}	0.156 ^{b}	0.0169	<0.001	0.512	0.968
Lag (h)	1.59 ^{ab}	1.68 ^{a}	1.36 ^{c}	1.49 ^{bc}	0.241	<0.002	0.034	0.778
NDFD (%)								
6 h	38.5 ^{a}	36.9 ^{ab}	36.0 ^{b}	34.7 ^{b}	2.32	0.009	0.084	0.810
12 h	60.8 ^{a}	60.4 ^{ab}	59.1 ^{b}	59.7 ^{ab}	0.80	0.030	0.889	0.301
24 h	72.7 ^{b}	72.3 ^{b}	74.7 ^{a}	76.7 ^{a}	0.74	<0.001	0.225	0.099
48 h	82.9 ^{b}	80.2 ^{c}	84.9 ^{a}	86.2 ^{a}	0.77	<0.001	0.196	<0.001
Total VFA (mmol/L)								
6 h	63.6 ^{ab}	63.9 ^{a}	60.0 ^{bc}	59.1 ^{c}	1.43	0.004	0.866	0.667
12 h	81.8 ^{a}	79.8 ^{ab}	76.1 ^{c}	78.0 ^{bc}	1.35	0.005	0.995	0.124
24 h	81.9	86.2	86.0	86.7	5.80	0.684	0.660	0.775
48 h	102.2	102.4	101.7	99.0	7.61	0.635	0.743	0.715

https://doi.org/10.3390/fermentation12010029

Table 3. *Cont.*

Item	Treatments [1]					p-Value [3]		
	C	E	I	EI	SEM [2]	I	E	E × I
Acetate (%)								
6 h	55.7	56.2	55.6	55.4	0.68	0.498	0.785	0.585
12 h	57.5	57.8	56.9	57.2	0.66	0.349	0.635	0.940
24 h	55.9	56.1	55.3	55.5	0.45	0.159	0.682	0.993
48 h	57.5	57.9	57.8	57.6	0.65	0.978	0.877	0.563
Propionate (%)								
6 h	27.4	27.1	27.7	27.4	1.39	0.798	0.838	0.970
12 h	25.4	25.0	26.3	26.0	1.23	0.425	0.772	0.924
24 h	23.3	24.3	25.1	24.9	1.00	0.252	0.673	0.548
48 h	24.0	23.6	23.8	24.1	1.13	0.893	0.954	0.703
Butyrate (%)								
6 h	14.0	13.8	13.6	14.0	0.85	0.899	0.929	0.728
12 h	13.7	13.8	13.2	13.3	0.70	0.457	0.931	0.950
24 h	14.9	13.8	13.8	13.8	0.89	0.518	0.535	0.547
48 h	12.5	12.6	12.5	12,4	0.63	0.874	0.982	0.850
Iso-butyrate								
6 h	0.60	0.62	0.63	0.65	0.023	0.108	0.258	0.974
12 h	0.67	0.68	0.68	0.68	0.010	0.360	0.923	0.360
24 h	1.13	1.13	1.14	1.13	0.025	0.957	0.713	0.753
48 h	1.27	1.27	1.22	1.22	0.029	0.042	0.882	0.984
Valerate (%)								
6 h	1.49	1.47	1.57	1.57	0.128	0.447	0.995	0.950
12 h	1.48	1.47	1.57	1.58	0.108	0.355	0.960	0.929
24 h	1.97	2.00	1.99	2.03	0.078	0.692	0.566	0.941
48 h	1.92	1.87	1.83	1.90	0.104	0.732	0.964	0.508
Iso-valerate (%)								
6 h	0.71 [b]	0.69 [b]	0.78 [a]	0.80 [a]	0.086	<0.001	0.834	0.192
12 h	0.96 [bc]	0.95 [c]	1.02 [a]	1.00 [ab]	0.073	0.001	0.409	0.631
24 h	2.22	2.16	2.25	2.21	0.124	0.552	0.506	0.898
48 h	2.42	2.42	2.41	2.43	0.075	0.921	0.780	0.716
Acetate/Propionate								
6 h	2.11	2.15	2.06	2.07	0.131	0.617	0.851	0.890
12 h	2.32	2.38	2.22	2.24	0.132	0.339	0.737	0.907
24 h	2.44	2.35	2.23	2.26	0.103	0.137	0.751	0.559
48 h	2.45	2.51	2.48	2.43	0.143	0.788	0.963	0.706

[1] Treatments included C, uninoculated control silage; E, corn silage ensiled with 0.5 mg/kg fresh forage chitinase; I, corn silage inoculated (cfu/g fresh forage) with lactic acid bacterial (LAB) inoculant of 1.5×10^8 *L. buchneri* and 1.5×10^8 *L. hilgardii* for a total of 3.0×10^8 LAB; EI, corn silage inoculated with combination of E and I. [2] SEM, pooled standard error of mean ($n = 3$). [3] p-values are for the effect of treatments having LAB inoculant (I and EI), chitinase enzyme (E and EI), and the interaction between LAB inoculant and chitinase (I × E). [4] Gas parameters were obtained by fitting gas production data to the equation $G = A \times [1 - e^{-c(t-Lag)}]$, where G is the cumulative gas production (in mL), A is the asymptotic gas production (in mL g^{-1} DM), c is the fractional fermentation rate (in h^{-1}), Lag is the initial delay in the onset of gas production (in h), and t is the gas reading time (in h). [a,b,c] within a row means without a common letter differ ($p < 0.05$) among treatments.

Disappearance of NDF was calculated at each sampling time point due to a treatment × time interaction ($p < 0.001$). No E × I interaction was observed except after 48 h, where the combination of E and I increased ($p < 0.001$) NDFD, as compared to C and E. Inoculant decreased ($p < 0.05$) NDFD at 6 and 12 h, but increased ($p < 0.001$) it at 24 and 48 h of the incubation. Treatments did not affect the total VFA production or molar proportions of individual VFA, except that I decreased ($p < 0.01$) total VFA production, but increased ($p < 0.001$) molar proportion of iso-valerate at 6 and 12 h of incubation.

 https://doi.org/10.3390/fermentation12010029

4. Discussion

4.1. Effects of Bacterial Inoculant and Chitinase on Ensiling Characteristics

Chitinases are glycosyl hydrolases present in a wide range of organisms from bacteria to humans that exhibit anti-fungal and anti-yeast activity by degrading the chitin layer within fungal cell walls [49–51]. These enzymes are gaining importance for their biotechnological applications, especially in pathogen control in human health and crop production [51]. Jin et al. [27] and Nair et al. [28] reported that application of *P. pentosaceus* with a chitinase to high-moisture alfalfa hay decreased internal temperature of baled forage. To our knowledge, this is the first study to explore the use of chitinase in corn silages.

The pH of the silages decreased rapidly in the initial phase of the ensiling, which was below 4.0 after 7 d ensiling, indicating that all silages were adequately ensiled regardless of treatment. It is generally regarded that spoilage microbial activity is inhibited when silage pH declines below 4.0, thereby preserving forage nutrients [4,52]. Although EI slightly increased the pH of terminal silages, the effect of this increase on silage quality would be minimal as the pH of these silages still remained below 4.0. This is supported by the similar OM, CP, starch, and NDF contents and similar DM loss for all silages. The decrease in silage pH is caused by the accumulation of LA and VFA from the metabolic activity of epiphytic microorganisms and the inoculated LAB utilizing the WSC as a substrate during ensiling. This is consistent with the lower residual WSC concentration in terminal silages as compared with that in freshly harvested corn forage and the accumulation of VFA and LA during the ensiling process. It should be noted that the residual WSC concentration was lower for I and EI than for C and E, indicating increased utilization of WSC as a result of inoculation with LAB. A significant reduction in residual WSC concentrations for terminal silages inoculated with heterofermentative LAB inoculants has been reported previously [6,11,18,31]. It is logical to assume that the LAB (*L. hilgardii* and *L. buchneri*) in the inoculated silages (I and EI) resulted in increased microbial activity and metabolism during the ensiling period, resulting in the utilization of WSC, as evidenced by the production of AC and VFA concentrations for I and EI than C and E (Table 1). Furthermore, both strains of the inoculant could produce LA and convert LA to AC [53–55]. This partially explained that C and I had similar levels of LA, but I had higher concentrations of AC, leading to a lower LA: AC ratio.

The increased acetate concentration after 21 d ensiling for I and EI indicated that the LAB inoculant increased acetate production at an early stage of ensiling. The inoculant used in this study contained both *L. buchneri* and *L. hilgardii* strains, which are heterofermentative and are capable of producing AC. Previous studies have shown that *L. buchneri* requires a longer fermentation time to produce AC [13], whereas *L. hilgardii* is capable of increasing AC within 14–21 d of ensiling to levels that improve the AS of silages [18,56]. Acetate is a potent anti-yeast and mold agent as reflected by a decrease in yeast and mold counts after d 21 of ensiling. It has been suggested that an LA:AC ratio of >3 is indicative of a dominant homolactic fermentation [19]. The LA:AC ratio in this study ranged from 1.6 to 3.9, with C and E silages having higher than 3 and I and IE silages being lower than 3. Other studies also observed that heterofermentative inoculants reduced the LA:AC ratio [6,12,16]. In contrast, the relatively lower LA:AC ratio of E compared to C silages was due to the lower LA concentration.

It appears that the combination of chitinase with *L. buchneri* and *L. hilgardii* did not negatively impact the activity of the LAB inoculant, as the AC concentration of silages treated with EI was similar to that of I. In contrast, in an evaluation of the effects of bacterial inoculants producing chitinase on corn silage, Joo et al. [57] reported no effect of a combination of *L. paracasei* L9-3 and *L. buchneri* L11-1 at 2.1×10^5 cfu/g forage DM

https://doi.org/10.3390/fermentation12010029

on AC or LA concentrations after 90 d or ensiling. In the present study, the reason why the LA concentration was decreased for E, despite having a similar LAB population as C, is unclear. Studies evaluating the effects of combining chitinase and LAB inoculants for forage preservation are scarce. Some recent studies evaluated the impact of chitosan alone or in combination with LAB inoculants on the ensiling of sugarcane, whole-crop soybean, and rehydrated corn kernels [58–63]. Chitosan is a biopolymer derived from chitin deacetylation and has been utilized as a silage inoculant due to its antimicrobial properties. Gandra et al. [62] added chitosan (5 g/kg fresh forage) in combination with 4×10^{10} cfu/g *L. plantarum* and 2.6×10^{10} cfu/g *Propionibacterium acidipropioni* to ensile whole plant soybean and recorded higher counts of TB, intermediate concentrations of AC, and lower concentrations of propionate, butyrate, iso-butyrate, and valerate, as compared to control forage. Results of the impact of application of chitosan alone have been mixed, with reports of increased TB counts [60,61], greater AC [59,61] and LA [61], decreased AC [60] and LA [58], or similar AC and LA concentrations [63] in sugar cane, rehydrated corn kernel or whole-crop soybean silages relative to untreated control silages. The variability in the experimental methodology, including the nature and concentrations of chitosan and the type of forage used between studies, could impact the end products of ensiling.

4.2. Effects of Bacterial Inoculant and Chitinase on Aerobic Stability of Whole-Crop Corn Silages

A significant result of this study was that LAB inoculant and the combination of LAB inoculant and chitinase, but not chitinase alone, markedly increased the AS of the silage from 152 h to >336 h. Silages are prone to spoilage upon exposure to air due to the activity of spoilage microorganisms that metabolize substrates, producing spores and toxins [19,64,65]. It has been observed that AS of silages is directly related to the presence of yeasts, with a population higher than 10^5 cfu/g DM being indicative of aerobic deterioration [12,66,67]. The yeast population in I and EI silages was lower than 10^5 cfu/g DM, while that in the E and C silages exceeded it, reflecting the improved AS in I and EI terminal silages as a result of increased AC concentrations. The decrease in AC concentrations and increase in yeast populations during 14 d of aerobic exposure for C and E silages occurred concomitantly, while the high AC concentration and low yeast population in I and EI silages remained relatively stable. Similar observations were also made by others evaluating combinations of heterofermentative LAB, including *L. buchneri* and *L. hilgardii*, in inoculated corn silages [6,12,16,29,55,56]. In contrast, Joo et al. [57] reported a decrease in aerobic stability of corn silages with bacterial inoculants producing chitinase, where the inoculants did not increase the concentrations of AC or decrease the numbers of yeast and mold. Studies evaluating the impact of the application of chitosan on aerobic stability have also been mixed. Del Valle et al. [58] reported that chitosan alone, but not in combination with a LAB, increased the aerobic stability of sugarcane silages, whereas Gandra et al. [62] reported that whole plant soybean silage treated with chitinase and LAB maintained a pH below 5 until 72 h after silo opening. The strains of LAB used, the nature of the studies, and the ensiling conditions could potentially affect the fermentation and microbiological outcomes of these studies.

The results showed that the application of chitinase did not adversely affect the fermentation or AS of whole-crop corn silage. The relatively greater TB and LAB counts, as well as higher concentrations of AC and LA for EI than for E, likely indicate that the addition of chitinase did not alter the microbial population and metabolism, and thereby the ensiling fermentation of corn silages. Another contributing factor could be the low concentration of mold in the silages, which was undetectable after 7 d of ensiling. Similar observations have been reported previously, where mold populations were low or undetectable in silages

https://doi.org/10.3390/fermentation12010029

during ensiling [6,16,31]. It should also be noted that the present study was carried out using minisilos under experimental conditions. Fermentation and microbial dynamics in commercial large-scale silos may differ from those in laboratory-scale minisilos, as ensiling and feedout conditions are complex and challenging. However, previous studies from our lab indicated similar trends between fermentation parameters and microbial populations during ensiling and aerobic exposure in both laboratory-scale minisilos and commercial-scale silobag studies [17]. Further study is needed to explore the potential use of chitinase as a biological control method in forage conservation.

4.3. Effects of Bacterial Inoculant and Chitinase on In Vitro Ruminal Fermentation of Silage

The present study showed that GP from 6 h to the end of the 48 h incubation was lower for I and EI than for C and E. This is likely due to the fact that the WSC concentrations were lower for I and EI. Water-soluble carbohydrates are simple sugars that are a readily available substrate for fermentation microbes. Therefore, the lower WSC concentrations for I and EI likely reduced the initial fermentation activity and corresponding GP. This was corroborated by the observation that fermentation of I and EI silages had lower total VFA than C and E silages during the first 12 h incubation. An interesting finding of this study was that LAB alone increased silage NDFD, which was further increased by inoculant and chitinase at 24 and 48 h of the in vitro incubation. Information about the effects of the combination of *L. hilgardii* and *L. buchneri* on ruminal digestion of corn silage is scarce. Tabacco et al. [67] reported that corn silage inoculated with *L. buchneri* in laboratory-scale plastic jars had higher 24 and 48 h in vitro NDFD than uninoculated corn silages. Arriola et al. [13] also showed that *L. hilgardii* and *L. buchneri*, either alone or in combination, increased in vitro NDFD of corn silages. Our previous study found, however, that beef cattle fed a diet containing 65% corn silage with and without a mixture of *L. hilgardii* and *L. buchneri* in large-scale silo bags had similar total tract nutrient digestibility [6]. The discrepancy among studies in the effect of heterolactic LAB on nutrient digestibility may reflect the difference in the strains of LAB used, the nature of studies (in vitro vs. in vivo), and ensiling conditions (mini-silo vs. silo-bag). For example, Nair et al. [6,17] reported no difference in the growth performance of beef cattle fed corn silages inoculated with or without a mixture of *L. hilgardii* and *L. buchneri* and ensiled in silo-bags. These authors concluded that feeding inoculated silages with enhanced AS did not result in improved animal growth performance compared to uninoculated silages, as the uninoculated silages were aerobically stable when fed (~3–4 d of AE). These authors also reported that the impact of inoculation could be more prominent in large, commercial-scale feeding operations where silages at the silo face may be exposed to air for longer periods of time, compromising the AS and nutritive value.

There was a LAB and chitinase interaction for NDFD, where the NDFD tended ($p = 0.099$) to be greater for I than for E after 24 h of incubation, while the NDFD was greater ($p < 0.001$) for I than for E after 48 h, even though the application of chitinase alone did not affect NDFD. Our previous studies also found that the combination of *P. pentosaceus* with chitinase increased the ruminal NDF degradability of high-moisture alfalfa hay [27,28]. Therefore, further research is needed to elucidate the mode of action of this interaction to further explore the potential of chitinase in forage conservation.

5. Conclusions

Application of *L. hilgardii* and *L. buchneri* reduced the number of yeasts, increased AC and total VFA concentrations in terminal silages, enhanced the aerobic stability, and improved NDF ruminal digestion of corn silages. Application of chitinase alone had

https://doi.org/10.3390/fermentation12010029

no adverse effect on ensiling fermentation, aerobic stability, or in vitro fermentation of whole-crop corn silages. Furthermore, combining the LAB inoculant and chitinase likely resulted in a synergistic effect, as indicated by greater TB and LAB counts, and suggested by a numerical improvement in 24 and 48 h NDFD. Further metagenomic evaluation during the ensiling and AE periods is necessary to define the microbial and fermentation characteristics of corn silages inoculated with chitinase for the potential application as a biological control in forage preservation.

Author Contributions: Conceptualization, T.A.M., E.C. and Y.W.; methodology, T.A.M., E.C. and Y.W.; software, H.N., J.N. and Y.W.; validation, T.A.M. and Y.W.; formal analysis, H.N., H.-E.Y., J.N. and Y.W.; investigation, H.N., H.-E.Y. and J.N.; resources, T.A.M., E.C. and Y.W.; data curation, H.N., H.-E.Y., J.N. and Y.W.; writing—original draft preparation, H.N., J.N., H.-E.Y., T.A.M. and Y.W.; writing—review and editing, H.N., J.N., H.-E.Y., T.A.M., E.C. and Y.W.; visualization, H.N., J.N., H.-E.Y., T.A.M. and Y.W.; supervision, T.A.M. and Y.W.; project administration, T.A.M. and Y.W.; funding acquisition, T.A.M., E.C. and Y.W. All authors have read and agreed to the published version of the manuscript.

Funding: This research was supported by Lallemand Specialties Inc. (grant number J-002039), the Program for Young Talents of Science and Technology of the Universities of Inner Mongolia Autonomous Region [grant number NJYT22054], and the Inner Mongolia Science and Technology Project [grant numbers 2025YFDZ0123, 2023YFDZ0079, 2025ZY0111].

Institutional Review Board Statement: The animal study protocol was approved by the Institutional Review Board (or Ethics Committee) of Lethbridge Research and Development Center (protocol code 1830, dated 11 January 2019) for studies involving animals.

Informed Consent Statement: Not applicable.

Data Availability Statement: The original contributions presented in this study are included in the article. Further inquiries can be directed to the corresponding author.

Conflicts of Interest: The author Eric Chevaux was employed by the company Lallemand SAS, an affiliate of Lallemand Inc. The remaining authors declare that the research was conducted in the absence of any commercial or financial relationships that could be construed as potential conflicts of interest. The authors declare that this study received funding from Lallemand Specialties Inc. (grant number J-002039), the Program for Young Talents of Science and Technology of the Universities of Inner Mongolia Autonomous Region [grant number NJYT22054], and the Inner Mongolia Science and Technology Project [grant numbers 2025YFDZ0123, 2023YFDZ0079, 2025ZY0111]. The funder had the following involvement with the study: conceptualization, resources, funding acquisition, and writing—review and editing.

References

1. Erdman, R.A.; Piperova, L.S.; Kohn, R.A. Corn silage versus corn silage:alfalfa hay mixtures for dairy cows: Effects of dietary potassium, calcium, and cation-anion difference. *J. Dairy Sci.* **2011**, *94*, 5105–5110. [CrossRef] [PubMed]
2. Donna, M.A.; Roy, N.; Lee, C.; Lehmkuhler, J. Practical Corn Silage Harvest and Storage Guide for Cattle Producers. University of Kentucky Cooperative Extension Service. 2023. Available online: https://publications.mgcafe.uky.edu/sites/publications.ca.uky.edu/files/ID275.pdf (accessed on 17 December 2025).
3. Addah, W.; Baah, J.; Okine, E.K.; Owens, F.N.; McAllister, T.A. Effects of chop-length and ferulic acid esterase producing inoculant on fermentation and aerobic stability of barley silage, and growth performance of finishing feedlot steers. *Anim. Feed Sci. Technol.* **2014**, *197*, 34–46. [CrossRef]
4. Muck, R.E.; Nadeau, E.M.G.; McAllister, T.A.; Contreras-Govea, F.E.; Santos, M.C.; Kung, L., Jr. Silage review: Recent advances and future uses of silage additives. *J. Dairy Sci.* **2018**, *101*, 3980–4000. [CrossRef] [PubMed]
5. Puntillo, M.; Peralta, G.H.; Bürgi, M.D.M.; Huber, P.; Gaggiotti, M.; Binetti, A.G.; Vinderola, G. Meta-profiling of the bacterial community in sorghum silages inoculated with lactic acid bacteria. *J. Appl. Microbiol.* **2022**, *133*, 2375–2389. [CrossRef]

6. Nair, J.; Niu, H.; Andrada, E.; Yang, H.-E.; Chevaux, E.; Drouin, P.; McAllister, T.A.; Wang, Y. Effects of inoculation of corn silage with *Lactobacillus hilgardii* and *Lactobacillus buchneri* on silage quality, aerobic stability, nutrient digestibility, and growth performance of growing beef cattle. *J. Anim. Sci.* **2020**, *98*, Skaa267. [CrossRef]
7. Driehuis, F.; Elferink, S.J.; Spoelstra, S.F. Anaerobic lactic acid degradation during ensilage of whole crop maize inoculated with Lactobacillus buchneri inhibits yeast growth and improves aerobic stability. *J. Appl. Microbiol.* **1999**, *87*, 583–594. [CrossRef]
8. Drouin, P.; Tremblay, J.; Chaucheyras-Durand, F. Dynamic succession of microbiota during ensiling of whole plant corn following inoculation with Lactobacillus buchneri and Lactobacillus hilgardii alone or in combination. *Microorganisms* **2019**, *7*, 595. [CrossRef]
9. Holzer, M.; Mayrhuber, E.; Danner, H.; Braun, R. The role of *Lactobacillus buchneri* in forage preservation. *Trends Biotechnol.* **2003**, *21*, 282–287. [CrossRef]
10. Killerby, M.A.; Almeida, S.T.R.; Hollandsworth, R.; Guimaraes, B.C.; Leon-Tinoco, A.; Perkins, L.B.; Henry, D.; Schwartz, T.J.; Romero, J.J. Effect of chemical and biological preservatives and ensiling stage on the dry matter loss, nutritional value, microbial counts, and ruminal in vitro gas production kinetics of wet brewer's grain silage. *J. Anim. Sci.* **2022**, *100*, skac095. [CrossRef]
11. Filya, I.; Sucu, E.; Karabulut, A. The effect of *Lactobacillus buchneri* on the fermentation, aerobic stability and ruminal degradability of maize silage. *J. Appl. Microbiol.* **2006**, *101*, 1216–1223. [CrossRef]
12. Kleinschmit, D.H.; Kung, L., Jr. A meta-analysis of the effects of *Lactobacillus buchneri* on the fermentation and aerobic stability of corn and grass and small-grain silages. *J. Dairy Sci.* **2006**, *89*, 4005–4013. [CrossRef] [PubMed]
13. Arriola, K.G.; Oliveira, A.S.; Jiang, Y.; Kim, D.; Silva, H.M.; Kim, S.C.; Amaro, F.X.; Ogunade, I.M.; Sultana, H.; Pech Cervantes, A.A.; et al. Meta-analysis of effects of inoculation with *Lactobacillus buchneri*, with or without other bacteria, on silage fermentation, aerobic stability, and performance of dairy cows. *J. Dairy Sci.* **2021**, *104*, 7653–7670. [CrossRef] [PubMed]
14. Wang, T.; Teng, K.; Cao, Y.; Shi, W.; Xuan, Z.; Zhou, J.; Zhang, J.; Zhong, J. Effects of *Lactobacillus hilgardii* 60TS-2, with or without homofermentative *Lactobacillus plantarum* B90, on the aerobic stability, fermentation quality and microbial community dynamics in sugarcane top silage. *Bioresour. Technol.* **2020**, *312*, 123600. [CrossRef] [PubMed]
15. da Silva, É.B.; Costa, D.M.; Santos, E.M.; Moyer, K.; Hellings, E.; Kung, L., Jr. The effects of *Lactobacillus hilgardii* 4785 and *Lactobacillus buchneri* 40788 on the microbiome, fermentation, and aerobic stability of corn silage ensiled for various times. *J. Dairy Sci.* **2021**, *104*, 10678–10698. [CrossRef]
16. da Silva, É.B.; Smith, M.L.; Savage, R.M.; Polukis, S.A.; Drouin, P.; Kung, L., Jr. Effects of *Lactobacillus hilgardii* 4785 and *Lactobacillus buchneri* 40788 on the bacterial community, fermentation and aerobic stability of high-moisture corn silage. *J. Appl. Microbiol.* **2021**, *130*, 1481–1493. [CrossRef]
17. Nair, J.; Yang, H.-E.; Redman, A.-A.; Chevaux, E.; Drouin, P.; McAllister, T.A.; Wang, Y. Effects of a mixture of *Lentilactobacillus hilgardii, Lentilactobacillus buchneri, Pediococcus pentosaceus* and fibrolytic enzymes on silage fermentation, aerobic stability and performance of growing beef cattle. *Tans. Anim. Sci.* **2022**, *6*, txac144. [CrossRef]
18. Carvalho, B.F.; Ávila, C.L.S.; Pinto, J.C.; Neri, J.; Schwan, R.F. Microbiological and chemical profile of sugar cane silage fermentation inoculated with wild strains of lactic acid bacteria. *Anim. Feed Sci. Technol.* **2014**, *195*, 1–13. [CrossRef]
19. Ferrero, F.; Piano, S.; Tabacco, E.; Borreani, G. Effects of conservation period and Lactobacillus hilgardii inoculum on the fermentation profile and aerobic stability of whole corn and sorghum silages. *J. Sci. Food Agric.* **2019**, *99*, 2530–2540. [CrossRef]
20. Hoster, F.; Schmitz, J.E.; Daniel, R. Enrichment of chitinolytic microorganisms: Isolation and characterization of a chitinase exhibiting antifungal activity against phytopathogenic fungi from a novel Streptomyces strain. *Appl. Microbiol. Biotechnol.* **2005**, *66*, 434–442. [CrossRef]
21. Oyeleye, A.; Normi, Y.M. Chitinase: Diversity, limitations, and trends in engineering for suitable applications. *Biosci. Rep.* **2018**, *38*, BSR2018032300. [CrossRef]
22. Pasin, T.M.; de Oliveira, T.B.; Scarcella, A.S.A.; Polizeli, M.L.T.M.; Guazzaroni, M.E. Perspectives on expanding the repertoire of novel microbial chitinases for biological control. *J. Agric. Food Chem.* **2021**, *69*, 3284–3288. [CrossRef]
23. Nagpure, A.; Choudhary, B.; Gupta, R.K. Chitinases: In agriculture and human healthcare. *Crit. Rev. Biotechnol.* **2014**, *34*, 215–232. [CrossRef] [PubMed]
24. Le, B.; Yang, S.H. Microbial chitinases: Properties, current state and biotechnological applications. *World J. Microbiol. Biotechnol.* **2019**, *35*, 144. [CrossRef] [PubMed]
25. Kozome, D.; Uechi, K.; Taira, T.; Fukada, H.; Kubota, T.; Ishikawa, K. Structural analysis and construction of a thermostable antifungal chitinase. *Appl. Environ. Microbiol.* **2022**, *88*, e0065222. [CrossRef] [PubMed]
26. Jin, L.; Duniere, L.; Lynch, J.P.; Chevaux, E.; McAllister, T.A.; Baah, J.; Wang, Y. Application of *Pediococcus pentosaceus, Pichia anomala*, and chitinase to high moisture alfalfa hay at baling: Effects on ruminal digestibility. In Proceedings of the ASAS Joint Annual Meeting 2015, Orlando, FL, USA, 9–12 July 2015.

https://doi.org/10.3390/fermentation12010029

27. Jin, L.; Chevaux, E.; McAllister, T.; Baah, J.; Drouin, P.; Wang, Y. Impact of *Pediococcus pentosaceus* and *Pichia anomala* in combinatio with chitinase on the preservation of high-moisture alfalfa hay. *Grass Forage Sci.* **2018**, *73*, 610–621. [CrossRef]
28. Nair, J.; Jin, L.; Chevaux, E.; McAllister, T.A.; Wang, Y. Impact of inoculation with *Pediococcus pentosaceus* in combination wit chitinase on bale core temperature, nutrient composition, microbial ecology, and ruminal digestion of high-moisture alfalfa ha *Fermentation* **2024**, *10*, 530. [CrossRef]
29. EFSA Panel on Additives and Products or Substances used in Animal Feed (FEEDAP); Bampidis, V.; Azimonti, G.; Bastos, M.L Christensen, H.; Dusemund, B.; Fašmon Durjava, M.; Kouba, M.; López-Alonso, M.; López Puente, S.; et al. Safety and efficac of a feed additive consisting of *Propionibacterium freudenreichii* DSM 33189 and *Lentilactobacillus buchneri* (formerly *Lactobacillu buchneri*) DSM 12856 for all animal species (Lactosan GmbH & Co.KG.). *EFSA J.* **2022**, *20*, e07151. [CrossRef]
30. Sigma-Aldrich, 2025. Available online: https://www.sigmaaldrich.com/US/en/product/sigma/c6137 (accessed on 20 Novem ber 2025).
31. Nair, J.; Xu, S.; Smiley, B.; Yang, H.-E.; McAllister, T.A.; Wang, Y. Effects of inoculation of corn silage with *Lactobacillus* spp or *Saccharomyces cerevisiae* alone or in combination on silage fermentation characteristics, nutrient digestibility, and growt performance of growing beef cattle. *J. Anim. Sci.* **2019**, *97*, 4974–4986. [CrossRef]
32. Teller, R.S.; Schmidt, R.J.; Whitlow, L.W.; Kung, L., Jr. Effect of physical damage to ears of corn before harvest and treatment wit various additives on the concentration of mycotoxins, silage fermentation and aerobic stability of corn silage. *J. Dairy Sci.* **2012** *95*, 1428–1436. [CrossRef]
33. National Academies of Sciences, Engineering, and Medicine (NASEM). *Nutrient Requirements of Beef Cattle*, 8th revised ed National Academies Press: Washington, DC, USA, 2016.
34. Canadian Council on Animal Care (CCAC). *CCAC Guidelines on: The Care and Use of Farm Animals in Research, Teaching, and Testing* CCAC: Ottawa, ON, Canada, 2009.
35. Menke, K.H.; Raab, L.; Salewski, A.; Steingass, H.; Fritz, D.; Schneider, W. The estimation of the digestibility and metabolizabl energy content of ruminant feeding stuffs from the gas production when they are incubated with rumen liquor in vitro. *J. Agric Sci.* **1979**, *93*, 217–222. [CrossRef]
36. Addah, W.; Baah, J.; Groenewegen, P.; Okine, E.K.; McAllister, T.A. Comparison of the fermentation characteristics, aerobi stability and nutritive value of barley and corn silages ensiled with or without a mixed bacterial inoculant. *Can. J. Anim. Sci.* **2011** *91*, 133–146. [CrossRef]
37. Nelson, N. A photometric adaptation of the Somogyi method for the determination of glucose. *J. Biol. Chem.* **1944**, *153*, 375–38 [CrossRef]
38. Kudo, H.; Cheng, K.-J.; Costerton, J.W. Interactions between *Treponema bryantii* and cellulolytic bacteria in the in vitro degradatio of straw cellulose. *Can. J. Microbiol.* **1987**, *33*, 244–248. [CrossRef] [PubMed]
39. Broderick, G.A.; Kang, J.H. Automated simultaneous determination of ammonia and total amino acids in ruminal fluid an in vitro media. *J. Dairy Sci.* **1980**, *63*, 64–75. [CrossRef]
40. Association of Official Analytical Chemists. *Official Methods of Analysis*, 17th ed.; AOAC International: Gaithersburg, MD, USA 2000.
41. Van Soest, P.J.; Robertson, J.B.; Lewis, B.A. Methods for dietary fiber, neutral detergent fiber, and nonstarch polysac-charides in relation to animal nutrition. *J. Dairy Sci.* **1991**, *74*, 3583–3597. [CrossRef]
42. Herrera-Saldana, G.E.; Huber, J.T.; Poore, M.H. Dry matter, crude protein, and starch degradability of five cereal grains. *J. Dairy Sci.* **1990**, *73*, 2386–2393. [CrossRef]
43. Addah, W.; Baah, J.; McAllister, T.A. Effects of an exogenous enzyme-containing inoculant on fermentation characteristics of barley silage and on growth performance of feedlot steers. *Can. J. Anim. Sci.* **2016**, *96*, 1–10. [CrossRef]
44. Snell, F.D.; Snell, C.T. *Colorimetric Methods of Analysis*; D. van Nostrand Company Inc.: Toronto, ON, Canada, 1953; p. 204.
45. Hristov, A.N.; McAllister, T.A.; Cheng, K.-J. Stability of exogenous polysaccharide-degrading enzyme in the rumen. *Proc. West. Sect. ASAS* **1997**, *48*, 145–148. [CrossRef]
46. Assis, F.G.V.; Ávila, C.L.S.; Pinto, J.C.; Schwan, R.F. New inoculants on maize silage fermentation. *R. Bras. Zootec.* **2014**, *43*, 395–403. [CrossRef]
47. Lopez, S.; France, J.; Dhanoa, M.S.; Mould, F.; Dijkstra, J. Comparison of mathematical models to describe disappearance curves obtained using the polyester bag technique for incubating feeds in the rumen. *J. Anim. Sci.* **1999**, *77*, 1875–1888. [CrossRef]
48. SAS Inst. Inc. *SAS OnlineDoc 9.3.1*; SAS Institute Inc.: Cary, NC, USA, 2012.
49. De Boer, W.; Klein Gunnewiek, P.J.; Lafeber, P.; Janse, J.D.; Spit, B.E.; Woldendorp, J.W. Anti-fungal properties of chitinolytic dune soil bacteria. *Soil Biol. Biochem.* **1998**, *30*, 193–203. [CrossRef]
50. Chang, W.; Chen, C.; Wang, S. An antifungal chitinase produced by bacillus cereus with shrimp and crab shell powder as a carbon source. *Curr. Microbiol.* **2003**, *47*, 102–108. [CrossRef] [PubMed]

 https://doi.org/10.3390/fermentation12010029

51. Singh, R.; Updhyay, S.K.; Singh, M.; Sharma, I.; Sharma, P.; Kamboj, P.; Saini, A.; Voraha, R.; Sharma, A.K.; Upadhyay, T.K.; et al. Chitin, chitinases and chitin derivatives in biopharmaceutical, agricultural, and environmental perspective. *Biointerface Res. Appl. Chem.* **2021**, *11*, 9985–10005. [CrossRef]
52. Rooke, J.A.; Hatfield, R.D. Biochemistry of ensiling. In *Silage Science and Technology (Agronomy Series No. 42)*; Buxton, D.R., Muck, R.E., Harrison, H.J., Eds.; American Society of Agronomy: Madison, WI, USA, 2003; pp. 95–139.
53. Oude Elferink, S.J.; Krooneman, J.; Gottschal, J.C.; Spoelstra, S.F.; Faber, F.; Driehuis, F. Anaerobic Conversion of Lactic Acid to Acetic Acid and 1,2-Propanediol by *Lactobacillus buchneri*. *Appl. Environ. Microbiol.* **2001**, *67*, 125–132. [CrossRef] [PubMed]
54. Ferrero, F.; Tabacco, E.; Piano, S.; Casale, M. Temperature during conservation in laboratory silos affects fermentation profile and aerobic stability of corn silage treated with *Lactobacillus buchneri*, *Lactobacillus hilgardii*, and their combination. *J. Dairy Sci.* **2021**, *104*, 1696–1713. [CrossRef]
55. Reich, L.J.; Kung, L., Jr. Effects of combining *Lactobacillus buchneri* 40788 with various lactic acid bacteria on the fermentation and aerobic stability of corn silage. *Anim. Feed Sci. Technol.* **2010**, *159*, 105–109. [CrossRef]
56. Reis, C.B.; Santos, A.d.O.d.; Carvalho, B.F.; Schwan, R.F.; Ávila, C.L.d.S. Wild *Lactobacillus hilgardii* (CCMA 0170) strain modifies the fermentation profile and aerobic stability of corn silage. *J. Appl. Anim. Res.* **2018**, *46*, 632–638. [CrossRef]
57. Joo, Y.H.; Jeong, S.M.; Kim, J.; Seo, M.J.; Baeg, C.H.; Lee, S.S.; Kang, B.S.; Lee, Y.; Kim, J.W.; Kim, S.-C. Application effects of bacterial inoculants producing chitinase on corn silage. *J. Korean Soc. Grassl. Forage Sci.* **2023**, *43*, 148–155. [CrossRef]
58. Del Valle, T.A.; Zenatti, T.F.; Antonio, G.; Campana, M.; Gandra, J.R.; Zilio, E.M.C.; de Mattos, L.F.A.; de Morais, J.G.P. Effect of chitosan on the preservation quality of sugarcane silage. *Grass Forage Sci.* **2018**, *73*, 630–638. [CrossRef]
59. Del Valle, T.A.; Antonio, G.; Zilio, E.M.C.; Dias, M.S.S.; Gandra, J.R.; Castro, F.A.B.; Campana, M.; Morais, J.P.G. Chitosan level effects on fermentation profile and chemical composition of sugarcane silage. *Braz. J. Vet. Res. Anim. Sci.* **2020**, *57*, e162942. [CrossRef]
60. Del Valle, T.A.; Morais, J.P.G.; Facco, F.B.; Gandra, J.R.; Campana, M.; Capucho, E.; Garcia, T.M. Chitosan decreases fermentation losses and improves aerobic stability of rehydrated corn silage. *Ciência Rural* **2025**, *55*, e20230164. [CrossRef]
61. Gandra, J.R.; Oliveira, E.R.; Takiya, C.S.; Goes, R.H.T.B.; Paiva, P.G.; Oliveira, K.M.P.; Gandra, E.R.S.; Orbach, N.D.; Haraki, H.M.C. Chitosan improves the chemical composition, microbial quality, and aerobic stability of sugarcane silage. *Anim. Feed Sci. Technol.* **2016**, *214*, 44–52. [CrossRef]
62. Gandra, J.R.; Takiya, C.S.; Del Valle, T.A.; Oliveira, E.R.; de Goes, R.H.T.B.; Gandra, E.R.S.; Batista, J.D.O.; Araki, H.M.C. Soybean whole-plant ensiled with chitosan and lactic acid bacteria: Microorganism counts, fermentative profile, and total losses. *J. Dairy Sci.* **2018**, *101*, 7871–7880. [CrossRef]
63. Morais, J.P.G.; Cantoria Junior, R.; Garcia, T.M.; Capucho, E.; Campana, M.; Gandra, J.R.; Ghizzi, L.G.; Del Valle, T.A. Chitosan and microbial inoculants in whole-plant soybean silage. *J. Agric. Sci.* **2021**, *159*, 227–235. [CrossRef]
64. Hu, Z.; Niu, H.; Tong, Q.; Chang, J.; Yu, J.; Li, S.; Zhang, S.; Ma, D. The microbiota dynamics of alfalfa silage during ensiling and after air exposure, and the metabolomics after air exposure are affected by *Lactobacillus casei* and cellulase addition. *Front. Microbiol.* **2020**, *11*, 519121. [CrossRef]
65. da Silva, B.É.; Liu, X.; Mellinger, C.; Gressley, T.F.; Stypinski, J.D.; Moyer, N.A.; Kung, L., Jr. Effect of dry matter content on the microbial community and on the effectiveness of a microbial inoculant to improve the aerobic stability of corn silage. *J. Dairy Sci.* **2022**, *105*, 5024–5043. [CrossRef]
66. Kung, L., Jr. Aerobic stability of silage. In Proceedings of the California Alfalfa & Forage Symposium and Corn/Cereal Silage Conference, 1–2 December 2010; University of California Cooperative Extension: Visalia, CA, USA; Davis, CA, USA, 2010. Available online: https://alfalfasymposium.ucdavis.edu/+symposium/proceedings/2010/10-89.pdf (accessed on 27 October 2025).
67. Tabacco, E.; Righi, F.; Quarantelli, A.; Borreani, G. Dry matter and nutritional losses during aerobic deterioration of corn and sorghum silages as influenced by different lactic acid bacteria inocula. *J. Dairy Sci.* **2011**, *94*, 1409–1419. [CrossRef]

 https://doi.org/10.3390/fermentation12010029

fermentation

MDPI

Article

Productivity, Fermentation Parameters, and Chemical Composition of Silages from Biomass Sorghum Hybrids in Ratoon Crop

Yara América da Silva [1], Marco Antonio Previdelli Orrico Junior [1,*], Marciana Retore [2], Gessí Ceccon [2], Isabele Paola de Oliveira Amaral [1], Ana Carolina Amorim Orrico [1], Giuliano Reis Pereira Muglia [1] and Tatiane Fernandes [3]

[1] College of Agricultural Science, Federal University of Grande Dourados, Dourados 79804-970, MS, Brazil; yaraamerica603@gmail.com (Y.A.d.S.); isabelep.oliveira@gmail.com (I.P.d.O.A.); anaorrico@ufgd.edu.br (A.C.A.O.); gmuglia12@gmail.com (G.R.P.M.)
[2] Brazilian Agricultural Research Corporation-Embrapa Agropecuária Oeste, Dourados 79804-970, MS, Brazil; marciana.retore@embrapa.br (M.R.); gessi.ceccon@embrapa.br (G.C.)
[3] Department of Dairy Science, Virginia Tech, Blacksburg, VA 24061-0315, USA; fernandest@vt.edu
* Correspondence: marcojunior@ufgd.edu.br; Tel.: +55-67-98116-0711

Abstract: Biomass sorghum stands out for its high dry matter yield and ratooning ability, enabling additional harvests and silage production. This study evaluated the productive potential and fermentation quality of silages from ratoon biomass sorghum hybrids. A 5×2 factorial randomized block design was used, with five hybrids (CMSXS5039, CMSXS5044, CMSXS7102, CMSXS7103, and BRS 716) grown in two municipalities of Mato Grosso do Sul, Brazil (Dourados and Jateí). Dry matter production (DMP) did not differ ($p > 0.05$) among the hybrids within each municipality; however, overall yield was higher in Jateí, averaging 12 t DM/ha. In Dourados, CMSXS5039 and CMSXS5044 showed the highest lactic acid concentrations (46.71 and 59.73 g/kg DM), whereas in Jateí, CMSXS7102, CMSXS7103, and BRS 716 stood out (45.70, 44.78, and 40.77 g/kg DM, respectively), among the sites, Jateí had the greater lactic acid production (49.95 g/kg DM). Aerobic stability (AS) averaged 28.5 h, with higher values in Dourados ($p < 0.05$), about 16 h longer than in Jateí. BRS 716 and CMSXS5044 presented the highest crude protein contents (115.17 and 118.33 g/kg DM). CMSXS5039 grown in Jateí had the lowest neutral detergent fiber and the highest starch and non-fiber carbohydrate values. Biomass sorghum hybrids exhibited good yield potential and good silage quality even under low rainfall conditions, with CMSXS5039 best suited for more energetic diets and BRS 716 and CMSXS5044 for more proteic diets.

Keywords: agronomic characteristics; ensiling; forage conservation; *Sorghum bicolor*

1. Introduction

Sorghum (*Sorghum bicolor* (L.) *Moench*) has stood out as a promising versatile crop in agricultural production given its adaptability to different climate and soil conditions, in addition to its efficient water use [1,2]. Among the several sorghum types, biomass sorghum has received special attention in the agricultural and livestock sector, particularly for silage production [3,4].

The choice of biomass sorghum for silage production is justified by its superior agronomic characteristics such as high biomass productivity, good nutritive value, and resistance to pests and diseases [5,6]. Furthermore, biomass sorghum has excellent capacity to adapt to more arid environments or seasons, which makes it a viable choice for regions where other forage crops (such as corn) lack satisfactory performance [7] according to the Agricultural Climate Risk Zoning.

A notable advantage of biomass sorghum is its capacity to ratoon after cutting, enabling two consecutive harvests within the same growing season [8]. Although the second harvest typically yields less biomass than the first, the combined production from both cuts can substantially increase total sorghum output per unit area, thereby reducing the cost per ton of biomass produced [8,9].

However, ratoon sorghum often exhibits marked differences in chemical composition compared with the first cut. Several studies have reported that ratooning tends to have lower dry matter concentrations and reduced levels of water-soluble carbohydrates, along with potential changes in fiber composition and lignification patterns [9]. These compositional shifts can impair the ensiling process by limiting the availability of fermentable substrates for lactic acid bacteria, potentially leading to slower pH decline, suboptimal fermentation profiles, and reduced aerobic stability (AS) of the resulting silage [10].

Given these constraints, there is a pressing need for breeding programs aimed at developing biomass sorghum hybrids that combine high ratoon yield with favorable chemical characteristics to support efficient fermentation and stable silage quality [11].

In face of that, the present research had the following hypotheses: (i) biomass sorghum hybrids sustain satisfactory ratoon yield, justifying an additional harvest; (ii) the silages of new hybrids CMSXS5039, CMSXS5044, CMSXS7102, and CMSXS7103 (in pre-launch phase) have superior nutritive value over silages produced from BRS 716 (commercial hybrid). Therefore, the present study aimed to assess the productive potential and fermentation quality of silages from ratoon biomass sorghum hybrids so as to identify which of those are the most promising for silage production.

2. Materials and Methods

The trial was simultaneously conducted in two regions of the state of Mato Grosso do Sul, Brazil. The first region was a farm located in the municipality of Jateí (22°26′57″ S, 54°20′11″ W, and 396 m above sea level), which has dystroferric Red Yellow Latosol—LVAd soil type according to the Brazilian Soil Classification System—SiBCS Ref. [12], equivalent to Typic Hapludox in the Soil Taxonomy Ref. [13], with 12% clay.

The second region was the experimental area of Brazilian Agricultural Research Corporation (Empresa Brasileira de Pesquisa Agropecuária—EMBRAPA) located in the municipality of Dourados (22°16′44″ S, 54°49′10″ W, and 430 m above sea level), which has eutrophic Red Latosol—Lve soil type according to the SiBCS Ref. [12], equivalent to Rhodic Eutrudox in the Soil Taxonomy Ref. [13], with 75% clay. The climate in both sites is Cwa (humid mesothermic, with wet summers) according to the Köppen classification [14]. Figure 1 shows the rain regime throughout the experimental period and the historical rainfall.

The experiment adopted a 5 × 2 factorial randomized block design comprising five biomass sorghum hybrids (CMSXS5039, CMSXS5044, CMSXS7102, CMSXS7103, and BRS 716) sowed in two distinct regions of the state of Mato Grosso do Sul, Brazil (Dourados and Jateí). Variation in soil fertility in the area was the blocking factor.

The soil analysis of the experimental area in Jateí had the following result: sand: 848 g/kg; silt: 32 g/kg; clay: 120 g/kg; pH $CaCl^2$: 4.7; Al^{3+}: 0.07 cmolc/dm^3; Ca^{2+}: 0.76 cmolc/dm^3; Mg^{2+}: 0.38 cmolc/dm^3; $H^+ + Al^{3+}$: 2.93 cmolc/dm^3; K^+: 0.80 cmolc/dm^3; P (Mehlich): 35.6 mg/dm^3; sum of bases: 1.94 (cmolc/dm^3); cation exchange capacity: 4.87 cmolc/dm^3; base saturation: 39.84%; and organic matter: 13.76 g/kg.

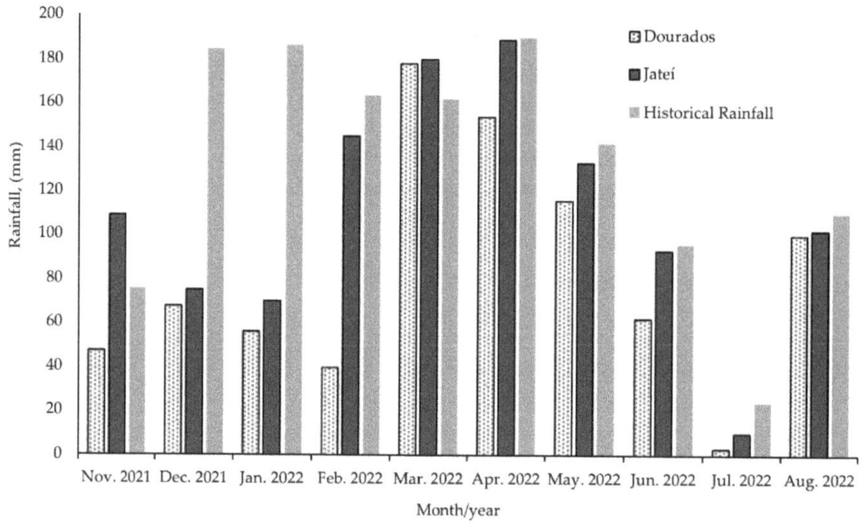

Figure 1. Rainfall (mm) recorded in Dourados and Jateí throughout the experimental period, and historical rainfall of the region (2002–2022).

The soil analysis of the experimental area in Dourados had the following result: sand: 205 g/kg; silt: 158 g/kg; clay: 637 g/kg; pH $CaCl^2$: 4.9; Al^{3+}: 0.07 cmolc/dm^3; Ca^{2+}: 3.10 cmolc/dm^3; Mg^{2+}: 1.36 cmolc/dm^3; $H^+ + Al^{3+}$: 4.15 cmolc/dm^3; $K^{+:}$ 0.64 cmolc/dm^3; P (Mehlich): 9.30 mg/dm^3; sum of bases: 5.10 cmolc/dm^3); cation exchange capacity: 9.25 cmolc/dm^3; base saturation: 55.14%; and organic matter: 27.89 g/kg. All physical and chemical soil analyses were carried out according to the methodology described by Ref. [15]

The hybrids were sown in the municipalities of Jateí and Dourados on 5 and 23 November 2021, respectively, using a SHM 1517 seed drill (Semeato Inc., Passo Fundo, RS, Brazil) with a row spacing of 45 cm. Fertilization was applied following recommendations based on soil chemical analysis and according to Ref. [16]. At planting, 300 kg/ha of fertilizer with an 8-20-20 (N-P-K) formula was applied. Subsequently, 20 days after plant emergence, a topdressing fertilization was carried out with 140 kg/ha of nitrogen in the form of ammonium sulfate. Sorghum density was set at 100,000 plants/ha, with an allowable margin of error of ±2.5%. Each experimental plot was composed of seven, 8 m rows. Six experimental plots were sowed per treatment.

The first sorghum harvest took place on 6 March 2022 in Jateí and on 12 April 2022 in Dourados. Fourteen days after cutting, top dressing at 140 kg/ha N, in the form of ammonium sulfate, was applied. The second harvest, after ratoon of the hybrids, took place on 19 July 2022 in Jateí and on 24 August 2022 in Dourados (Figure 2).

In order to assess the productive characteristics of the hybrids, a sample of 2 linear meters was collected from one of the central rows in each plot for morphological separation and determination of dry matter (DM) values of each plant part. The sorghum plants were split into leaf, stem with sheath, and panicle. Plant height was measured with a measuring tape and the stem diameter, with digital calipers. Forage yield per hectare was estimated by multiplying the total plant weight harvested in each plot by its respective content of DM. The results were expressed in tons of DM per hectare (t DM/ha).

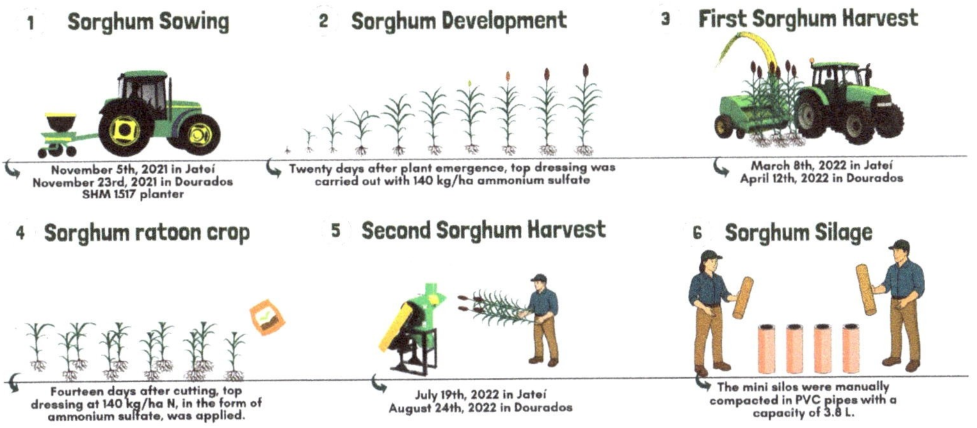

Figure 2. Representation of the production and harvest steps of the different biomass sorghum hybrids assessed in the municipalities of Jateí and Dourados, MS, Brazil.

For silage production, a sample of the central plants from each experimental plot was processed using a stationary forage grinder (JF, model 5D), which chopped the material to an average particle length of 1.5 cm. The material from each plot was homogenized and used to fill the experimental silos. During filling, forage samples from each treatment were collected and frozen at −18 °C for later analysis of chemical composition, pH, and buffer capacity. The ground forage mass was ensiled in laboratory experimental silos built using PVC pipes (10 cm diameter and 50 cm height) with useful volume of 3.8 L. The material was manually compacted using wood rods to achieve mean density of 270 kg DM/m^3. At the bottom of each experimental silo, 300 g sand was placed for effluent drainage, separated from the forage by a thin piece of fabric mesh. After filling, the experimental silos were sealed with double-faced (black and white) plastic film and adhesive tape and stored in the laboratory at room temperature.

To calculate fermentation losses, all silo components (silo, sand, and fabric), as well as the ensiled TMR mass, were weighed before and after ensiling. Dry matter recovery (g/kg of ensiled DM), gas losses (g/kg of ensiled DM), and effluent losses (g/kg of ensiled forage) were calculated according to the equations proposed by Ref. [17]. Dry matter recovery was calculated using the following formula:

$$\mathrm{DMR} = 100 - \left(\frac{\mathrm{DMI} - \mathrm{DMF}}{\mathrm{DMI}} \times 100\right)$$

where DMR = dry matter recovery (% of initial DM mass); DMI = initial DM mass (kg of DM placed in the silos); and DMF = final DM mass (kg of DM removed from the silos).

Gas losses were calculated using the following formula:

$$\mathrm{GL} = \frac{\mathrm{SWI} - \mathrm{SWF}}{\mathrm{DMI}} \times 100$$

where GL = gas loss during storage (% of initial DM mass); SWI = weight of the sealed silo at the beginning (kg); SWF = weight of the sealed silo at the time of opening (kg); and DMI = initial DM mass (kg of DM placed in the silos).

Effluent losses were calculated using the following formula:

$$\mathrm{EP} = \frac{\mathrm{SWF} - \mathrm{SWI}}{\mathrm{DMI}} \times 1000$$

where EP = effluent production (kg/t of DM); SWF = final weight of the set (silo + sand + fabric) in kg; SWI = initial weight of the set (silo + sand + fabric) in kg; and DMI = initial DM mass (kg of DM placed in the silos).

The silos were then opened and the material inside them was removed and homogenized for sample collection. The samples of each of the silages produced were sent to a partner laboratory for chemical composition analysis. Another portion of the samples was frozen for later pH analysis.

The chemical composition of both silage and fresh forage samples was determined using a Foss 5000 Transport near-infrared reflectance spectrophotometer (NIRS; Eden Prairie, MN, USA) with calibrations (WinISI version 4.6.11, FOSS Analytical A/S, Hillerød, Denmark) provided by the Dairy One Forage Laboratory (Ithaca, NY, USA). The analyses included determination of dry matter (DM), crude protein (CP), neutral detergent fiber (NDF), ether extract (EE), soluble carbohydrates (SC), non-fiber carbohydrates (NFC), and starch. Model performance was verified through cross-validation and, when applicable, external validation, considering the coefficient of determination (R^2), standard error of validation (SEV), and residual predictive deviation (RPD), with minimum acceptance criteria of $R^2 \geq 0.90$ and RPD ≥ 3.0.

A portion of the frozen samples was processed to obtain an aqueous extract, which was used to determine pH (in fresh forage and silage), buffering capacity (only in fresh forage), and organic acids (only in silage). For extract preparation, 25 g of forage were diluted in 225 mL of distilled water and manually homogenized for approximately 20 min. The pH of the extract was measured using a digital potentiometer (mPA210, MS Tecnopon), and buffering capacity was determined according to the method described by Ref. [18]. Acetic acid and butyric acid concentrations were quantified using a gas chromatograph with a mass spectrometry detector (GCMS QP 2010 Plus, Shimadzu, Kyoto, Japan) equipped with a capillary column (Stabilwax, Restek, Bellefonte, PA, USA; 60 m $\times$ 0.25 mm ID, 0.25 μm polyethylene cross-bond carbowax glycol). Lactic acid concentration was determined by the colorimetric method described by Ref. [19].

Aerobic stability was determined in all silages after the silos were opened. Samples (2 $\pm$ 0.005 kg) of each replicate of each treatment were freely placed in clean experimental silos. Temperature sensors were placed in the geometric center of the silages and a double layer of gauze was placed on top of each experimental silo to prevent drying and contamination while allowing for air penetration. Room temperature, as well as the temperature of each silage, was recorded every minute using a datalogger (RC-4, Elitech®, San Jose, CA, USA). Aerobic stability was defined as the number of hours the silage remained stable before rising by more than 2 °C above the ambient temperature [10].

The data were analyzed using the statistical program Sisvar 5.8 (Build 92). When the interaction of factors was significant ($\alpha \leq 0.05$), the factors were analyzed separately. In the case of non-significant interaction, the factors were analyzed by principal component analysis. The means were compared by Scott-Knott test at 5% level of significance. The data were analyzed according to the following model:

$$Yijk = \mu + \beta k + Si + SAj + S * SAij + \varepsilon ijk,$$

where Yijk = dependent variable, μ = overall average, βk = block effect (random effect; k = 1, 2, 3, 4, 5, and 6), Si = effect of different hybrids (fixed effect; i = CMSXS5039, CMSXS5044, CMSXS7102, CMSXS7103, and BRS 716), SAj = site effect (fixed effect; j = Dourados and Jateí), S * SAij = effect of the interaction between hybrids and sites, and εijk = random error associated with each observation.

3. Results

3.1. Productivity and Morphological Characteristics of the Biomass Sorghum Hybrids

No significant difference was observed for DM production (DMP) among the biomass sorghum hybrids. However, significant difference was found in DMP between the production sites, with Jateí having higher productions (Table 1).

Table 1. Agronomic characteristics and chemical composition of different biomass sorghum hybrids assessed in ratoon crops in the municipalities of Jateí and Dourados, MS, Brazil, 2022.

Parameters	(H)					(S)		SEM	*p*-Value		
	H5039	H5044	H7102	H7103	H716	Jateí	Dourados		H	S	H × S
				Agronomic characteristics							
DMP, t/ha	7.49	8.04	7.74	7.26	8.38	12.55	3.00	0.20	0.85	<0.01	0.708
HS, m	1.80 a	1.46 b	1.92 a	1.87 a	1.81 a	2.44	1.10	0.08	<0.01	<0.01	0.069
SD, mm	12.41 a	11.03 b	9.77 c	9.56 c	10.71 b	12.39	9.00	0.51	<0.01	<0.01	0.449
Stem, %	47.06	47.49	45.71	47.30	47.16	43.02	50.87	0.32	0.81	<0.01	0.181
Leaf, %	22.74 b	28.80 a	19.97 b	21.93 b	23.55 b	16.06	30.73	1.47	0.02	<0.01	0.227
Panicle, %	30.20	23.71	34.32	30.77	29.29	40.92	18.40	4.03	<0.01	<0.01	<0.01
				Chemical composition							
DM (g/kg FM)	320.04 b	314.23 b	356.18 a	370.44 a	324.52 b	397.26	264.11	10.21	<0.01	<0.01	0.054
CP (g/kg DM)	112.25 b	120.01 a	95.00 c	105.5 b	118.25 a	121.90	94.10	1.99	<0.01	<0.01	0.078
SC (g/kg DM)	68.25 b	73.50 b	45.75 c	62.00 b	92.00 a	75.10	57.50	9.75	0.04	0.07	0.128
ST (g/kg DM)	125.25 a	52.25 c	97 b	101.25 b	70.5 c	125.90	52.60	7.36	<0.01	<0.01	0.547
EE (g/kg DM)	23.750	27.50	21.50	22.75	27.25	30.90	18.20	1.93	0.18	<0.01	0.216
NFC (g/kg DM)	242.75	193.50	181.00	199.50	204.75	246.40	162.20	13.36	0.07	<0.01	0.402
NDF (g/kg DM)	594.25 b	642.02 a	669.75 a	648.25 a	622.75 b	577.00	693.80	12.03	0.01	<0.01	0.677

H = Hybrid; S = Site; H5039 = CMSXS5039; H5044 = CMSXS5044; H7102 = CMSX7102; H7103 = CMSXS7103; H716 = BRS716; HS = Plant height in meters; SD = Steam diameter in millimeters; DMP = total dry matter production; DM = dry matter; FM = fresh matter; CP = crude protein; NDF = neutral detergent fiber; ST = Starch; EE = ether extract; NFC = non-fiber carbohydrate; SC = soluble carbohydrate. SEM = standard error of the mean. Means followed by different letters differ according to Scott-Knott test at 5% probability.

Hybrid CMSXS5044 had the shortest mean height (1.46 m) and highest leaf proportion (28.8%) when compared with the other hybrids assessed. The sites also significantly impacted ($p < 0.01$) plant height and leaf proportion, with the plants grown in Jateí being taller (2.44 m) and having higher leaf proportion (30.73%). Stem diameter (SD) was influenced both by the hybrids and the sites ($p < 0.01$), with the largest SD found for CMSXS5039 and the largest mean SD found in the plants grown in Jateí.

A significant interaction ($p < 0.01$) was observed between the hybrids and sites regarding panicle proportion (Table 1 and Figure 3). Hybrids CMSXS5039 and CMSXS7102 stood out for having the highest panicle proportions ($p < 0.01$) when grown in Jateí.

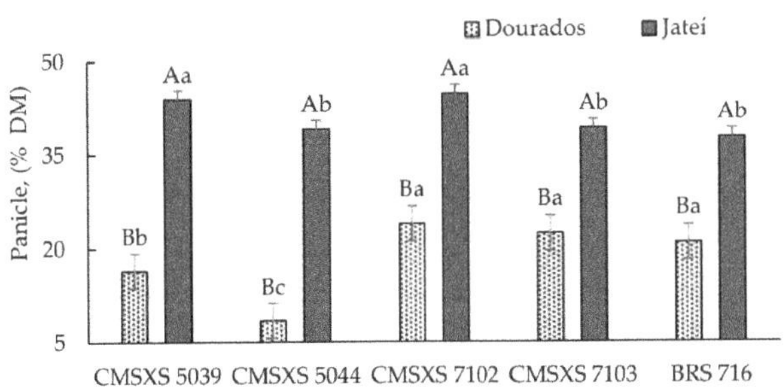

Figure 3. Panicle percentage in different biomass sorghum hybrids assessed in ratoon crops in the municipalities of Jateí and Dourados, MS, Brazil, 2022. Means followed by different letters differ according to Scott-Knott test at 5% probability. Capital letters compare sites and small letters compare hybrids.

Concerning chemical composition (Table 1), hybrids CMSXS7103 and CMSXS7102 had the highest DM values at 356.18 g/kg fresh matter (FM) and 370.44 g/kg FM, respectively, 15.17% higher than hybrid CMSXS5044, which had the lowest DM content (314.23 g/kg FM).

The values of CP, NDF, SC, and starch were influenced by both the hybrids and sites. Hybrid BRS 716 had the highest CP content, followed by hybrids CMSXS5039, CMSXS5044, CMSXS7103, and CMSXS7102, in that order. Hybrids BRS 716 and CMSXS5044 had the lowest starch values but stood out with the highest SC values. Hybrids CMSXS7102, CMSXS7103, and CMSXS5044 had the highest NDF values, whereas hybrids CMSXS5039 and BRS 716 had the lowest NDF values. On average, the plants grown in Jateí had the highest values of CP, starch, and SC and the lowest NDF values when compared with those grown in Dourados.

3.2. *Fermentation Parameters and Chemical Composition of Silages from Different Biomass Sorghum Hybrids*

A significant interaction ($p < 0.05$) was observed for EP between the hybrids and growth sites (Table 2 and Figure 4). Hybrids CMSXS5039 and CMSXS5044 grown in Dourados had the highest EP values ($p < 005$). However, no significant differences were observed for EP among silages produced with the different hybrids in Jateí. DMR was influenced exclusively by the hybrids, with CMSXS7102 having the lowest values (840.28 g/kg DM), approximately 3.56% lower than the average of the other hybrids. No differences ($p > 0.05$) were observed between hybrids and growth sites regarding GL. The average GL value found was 28.49 g/kg DM among the treatments (Table 2).

Table 2. Fermentation parameters and chemical composition of silages of different biomass sorghum hybrids assessed in ratoon crops in the municipalities of Jateí and Dourados, MS, Brazil, 2022.

Parameters	(H)					(S)		SEM	*p*-Value		
	H5039	H5044	H7102	H7103	H716	Jateí	Dourados		H	S	H × S
EP (kg/t DM)	59.66	51.75	46.13	44.06	41.23	4.39	92.74	4.62	<0.01	<0.01	0.022
DMR (g/kg DM)	858.74 a	871.33 a	840.28 b	860.96 a	860.73 a	859.64	857.18	4.31	0.04	0.688	0.136
GL (g/kg DM)	25.57	38.23	24.53	23.35	30.78	26.07	30.92	0.55	0.06	0.173	0.073
pH	3.84 b	3.82 b	3.91 a	3.80 c	3.77 c	3.91	3.74	0.07	<0.01	<0.01	0.065
Lac. Ac. (g/kg DM)	46.71	59.73	45.7	44.78	40.77	49.95	45.13	0.56	0.02	0.181	<0.01
Ace. Ac. (g/kg DM)	12.39	12.28	19.14	15.48	12.37	12.21	16.46	0.18	<0.01	0.01	<0.01
But. Ac. (g/kg DM)	1.40 a	0.43 c	1.29 b	1.44 a	0.88 b	1.04	1.14	0.02	<0.01	0.433	0.21
NH_3-N (g/kg TN)	9.48 a	5.01 c	9.47 a	7.92 a	6.71 b	9.12	6.47	0.08	<0.01	<0.01	0.31
AS (h)	25.55	31.98	28.29	28.9	27.75	20.78	36.71	3.61	0.51	<0.01	0.341
DM (g/kg FM)	301.00	307.92	351.08	340.34	311.42	399.6	245.1	9.79	<0.01	<0.01	<0.01
CP (g/kg DM)	113.50 b	118.33 a	109.75 c	110.58 c	115.17 a	128.57	98.37	2.98	0.04	<0.01	0.074
EE (g/kg DM)	38.25 b	46.5 a	38.17 b	37.25 b	36.58 b	44.76	32.2	1.86	0.05	<0.01	0.516
Starch (g/kg DM)	131.5	64.00	112.17	102	59.67	144.67	43.07	7.84	<0.01	<0.01	<0.01
NFC (g/kg DM)	279.83	228.5	235.33	241.5	229.58	270.43	215.47	1.85	<0.01	<0.01	<0.01
NDF (g/kg DM)	537.83	577.92	585.83	578.33	589.58	531.93	615.87	9.27	<0.01	<0.01	<0.01

H = Hybrid; S = Site; H5039 = CMSXS5039; H5044 = CMSXS5044; H7102 = CMSXS7102; H7103 = CMSXS7103; H716 = BRS716; EP = effluent production; DMR = dry matter recovery; GL = gas loss; pH = potential of hydrogen; Lac. Ac. = lactic acid; Ace. Ac. = acetic acid; But. Ac. = butyric acid; NH_3-N = ammoniacal nitrogen; AS = aerobic stability; DM = dry matter; FM = fresh matter; TN = total nitrogen; EE = ether extract; NFC = non-fiber carbohydrate; NDF = neutral detergent fiber; SEM = standard error of the mean. Means followed by different letters differ according to Scott-Knott test at 5% probability.

Hybrids CMSXS7103 and BRS 716 had the lowest ($p < 0.05$) pH values (3.80 and 3.77, respectively) while CMSXS7102 had the highest pH (3.91). Lactic acid values showed a significant interaction ($p < 0.05$) between the hybrids and sites. The silages produced by hybrids CMSXS5039 and CMSXS5044 in Dourados had the highest values of lactic acid (46.71 and 59.73 g/kg DM), whereas in Jateí, hybrids CMSXS7102, CMSXS7103, and BRS 716 produced silages with the highest lactic acid values (45.70, 44.78, and 40.77 g/kg DM, respectively).

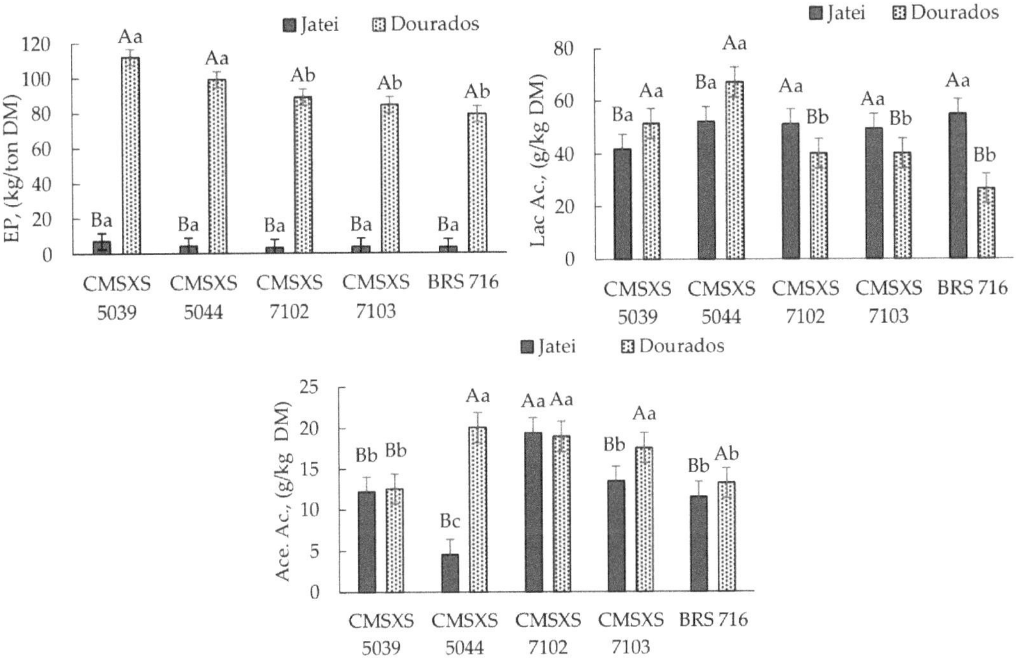

Figure 4. Effluent production (EP), lactic acid (Lac. Ac.) and acetic acid (Ace. Ac.) values of silages from different biomass sorghum hybrids assessed in ratoon crops in the municipalities of Jateí and Dourados, MS, Brazil, 2022. Means followed by different letters differ according to Scott-Knott test at 5% probability. Capital letters compare sites and small letters compare hybrids.

Overall, the hybrids grown in Dourados had the highest acetic acid values, except for CMSXS5039 and CMSXS7102, which had similar values in both sites. The highest acetic acid values were found in hybrids CMSXS7102 (Dourados and Jateí), CMSXS7103 (Dourados), and CMSXS5044 (Dourados), at an average of 18.95 g/kg DM. On the other hand, the lowest acetic acid production was found for hybrid CMSXS5044 grown in Jateí, at an average of 4.55 g/kg DM (Figure 4).

The lowest butyric acid and ammoniacal nitrogen (NH_3-N) values were found in hybrid CMSXS5044 (0.43 g/kg DM and 5.01 g/kg total nitrogen, respectively), with no significant difference between the sites (Table 2). No statistical difference was found in AS among the hybrids tested. However, the silages produced in Dourados had higher ($p < 0.05$) AS at 16 h longer on average when compared with those from Jateí.

Hybrids CMSXS7102 and CMSXS7103 had the highest DM values, while BRS 716 had the lowest DM content in Jateí. In Dourados, hybrid CMSXS5039 and CMSXS5044 had the lowest ($p < 0.05$) DM content (Figure 5). Overall, all hybrids grown in Dourados had lower DM values when compared with Jateí.

Crude protein also significantly varied ($p < 0.05$) among the hybrids tested. Hybrids BRS 716 and CMSXS5044 had the highest CP values at 115.17 g/kg DM and 118.33 g/kg DM, respectively. Hybrids CMSXS7102 and CMSXS7103, meanwhile, had the lowest CP values, approximately 7.25% lower than the highest values observed.

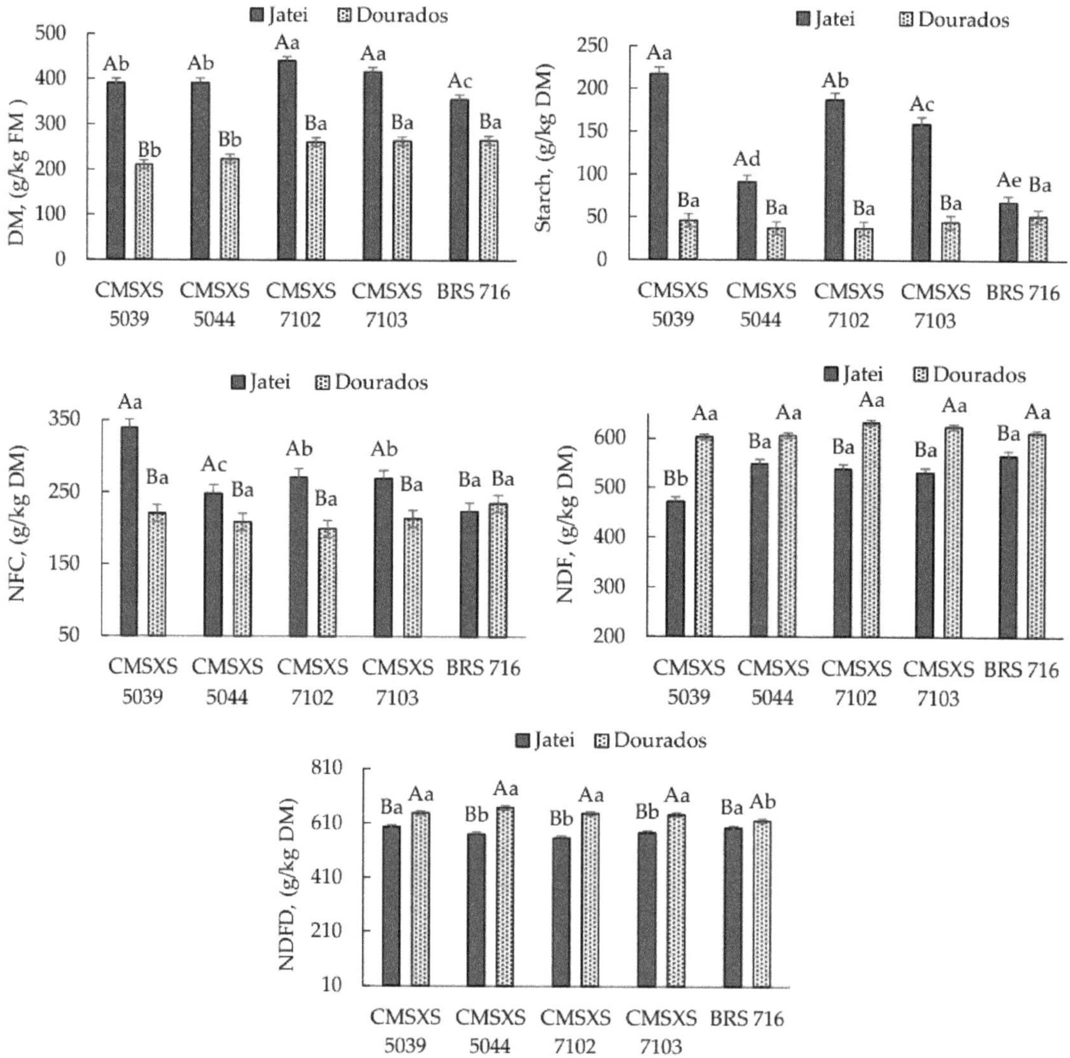

Figure 5. Contents of dry matter (DM), starch, non-fiber carbohydrates (NFC), neutral detergent fiber (NDF), and NDF digestibility (NDFD) of silages from different sorghum biomass hybrids assessed in regrowth in the municipalities of Jateí and Dourados, MS, Brazil. FM = fresh matter. Means followed by different letters differ according to Scott-Knott test at 5% probability. Capital letters compare sites and small letters compare hybrids.

Overall, the silages produced in Jateí had the lowest NDF values, with hybrid CMSXS5039 having the lowest NDF content among all treatments studied (537.83 g/kg DM). Ether extract content was higher only for hybrid CMSXS5044 (46.5 g/kg DM). No significant difference ($p > 0.05$) was found for starch or NFC values among the hybrids grown in Dourados. However, in Jateí, the silages produced from hybrid CMSXS5039 had the highest starch and NFC values.

4. Discussion

4.1. Productivity and Morphological Characteristics of the Biomass Sorghum Hybrids

The hybrid grown in Jateí achieved an average yield of 12.0 ± 2.3 t DM/ha, which can be considered high for the season. In contrast, the reduced rainfall in Dourados resulted in yields of only 3.0 ± 1.5 t DM/ha, indicating that, despite the recognized tolerance of sorghum hybrids to water stress, adequate minimum temperatures and soil moisture remain essential to achieve harvestable yields. In a study conducted by Ref. [20] evaluating four sorghum cultivars (SHS 400, 1G220, BRS 310, and 0992045) in semi-arid regions of Brazil, grain yields ranged from 1.7 to 5.1 t/ha. The authors reported a strong positive relationship between water availability and grain yield, with the highest productivity observed in areas where total precipitation during the crop cycle reached up to 519 mm.

According to Ref. [21], sorghum requires between 450 and 550 mm of water throughout its growth cycle, with optimal yields occurring when cumulative rainfall exceeds 500 mm. In the present study, the period from April to August—corresponding to the sorghum ratoon phase—recorded cumulative precipitation of approximately 430 mm in Dourados and 527 mm in Jateí, compared with a regional historical average of 562 mm for the same period (Figure 1—data from 2002 to 2022). These less favorable climatic conditions, particularly in Dourados, likely reduced yield potential and help explain the lower biomass production observed in this municipality.

Therefore, producers intending to use sorghum ratoon as a strategy to increase silage availability should consider anticipating planting to coincide with the onset of the rainy season, ensuring that ratoon occurs under still favorable temperature and precipitation conditions. Alternatively, harvesting at a lower DM content (below 33% DM) may also be an option to minimize yield losses.

Although they exhibited the same biomass productions, the tested biomass sorghum hybrids had different heights and leaf proportions. This suggests that while certain morphological parameters may be more stable, others, such as height and leaf proportion, are highly sensitive both to the genetic material and to the environment [22]. Hybrid CMSXS5044, which had the lowest height and highest leaf proportion, can be particularly advantageous in contexts where leaf biomass is preferred. On the other hand, plants grown in Jateí exhibited greater height and lowest leaf proportions, indicating that the climatic or edaphic conditions at this site are more favorable for vegetative growth.

Compared to other sorghum types, biomass sorghum has a longer vegetative cycle, which, combined with its greater plant height, generally results in higher productivity [2,11]. However, the total biomass per area is not influenced only by height Ref. [23], but also by stem diameter, number of shoots, population density, and photosynthesis efficiency of the plants [24]. That means a taller plant does not necessarily result in higher biomass production, thus explaining why in the present experiment hybrids of such different heights produced similar amounts of biomass.

A study by Ref. [11] showed that biomass sorghum varieties have an important correlation between plant height and fiber content in the plant. That is because very tall plants require a higher proportion of fiber for support so as to prevent losses due to lodging. However, Ref [2] did not find a significant correlation between the height of biomass sorghum hybrids and their fiber values. Such results suggest other factors such as hybrid genetics, growth conditions, and maturity stage may also impact the fiber content of biomass sorghum.

In the context of the present experiment, the plants grown in Dourados had significantly lower nutrient values than those grown in Jateí. An example of that difference is in starch values, where the plants from Jateí, which had higher proportions of panicles, also had the highest starch values.

According to Ref. [25], under stress conditions, plants tend to reduce the synthesis of sugars and proteins, decrease growth, store phenolic compounds, and increase tissue lignification, which may compromise forage nutritive value. That finding suggests the water stress in Dourados not only impacted the production of biomass sorghum hybrids but also compromised the nutritive value of the material produced.

Hybrids CMSXS5044 and BRS 716 stood out for their high CP values, attributed to the higher leaf proportions at harvest, a desirable characteristic to improve the nutritional value of the silage, particularly in diets of high-performance ruminants. Overall, the mean CP values of the hybrids were much higher than those commonly observed in the literature for biomass sorghum during the main harvest.

The study by Ref. [11] found CP values between 39 and 54 g/kg DM for biomass sorghum hybrids, which are much lower than the ones in the present experiment. It is likely that the low stature of the plants during ratoon contributed to the plants having lower fiber proportions and, consequently, higher protein values. In the experiment by Ref. [11], height ranged from 4 to 5.5 m and NDF values were approximately 700 g/kg DM.

4.2. *Fermentation and Nutritional Parameters of Silages from Biomass Sorghum Hybrids*

Effluent production during ensiling may be impacted by several factors that affect both its biomass accumulation and nutritive value. According to the literature, the main factors impacting EP are the DM content in the forage, the type of forage, compaction, particle size, the use of additives, pre-ensiling management, silo type, and climate conditions [26]. In the present experiment, DM content was the only factor that varied among the tested treatments and, therefore, had the greatest impact in the productions observed.

According to Ref. [26], forages with low DM content (below 30%) have more free water, which contributes to increasing effluent production and the loss of essential nutrients via leaching. That suggests the higher water content in the plants harvested in Dourados favored effluent generation Ref. [26], thus pointing to the need for using dry additives to adjust moisture and control effluent losses.

In contrast, the minimum GL values observed and the lack of significance among treatments (hybrids or site) are an indication of the efficiency of the fermentation process. According to Ref. [27], low GL in silages indicates lower DM losses during the fermentation process since gases are byproducts of microbial fermentation. Excess GL may indicate undesired fermentation and, consequently, significant losses of energy and nutrients [28]. Therefore, fermentation that produces less gas is more efficient and better preserves the quality of ensiled forage, which will often have higher lactic acid concentration in relation to other fatty acids [27].

In this study, the biomass sorghum silages obtained from ratoon crops showed lactic acid concentrations lower than those reported by Ref. [29] for the same hybrids in the main harvest. Nevertheless, the silages maintained a satisfactory fermentation profile Ref. [10], suggesting that the lactic acid produced was sufficient to ensure preservation and stability during storage.

Lactic acid plays a central role in silage quality, as it rapidly reduces the pH of the ensiled mass, creating an unfavorable environment for undesirable microorganisms such as *Clostridium* spp., which are associated with dry matter losses Ref. [27] and with undesirable fermentations, particularly butyric fermentation [10]. Moreover, once ingested by ruminants, lactic acid is converted into propionic acid, representing an important metabolic pathway for energy supply [10].

In general, acetic acid productions were low Ref. [10], particularly in the silages produced in Jateí, which suggests a prevalence of homofermentation (lactic acid only) over heterofermentation (lactic acid and other products, such as acetic acid) [30–32]. Studies have shown that sorghum undergoes significant changes in chemical composition during ratoon Ref. [9], which may result in changes in the fermentation profile of the silages.

A study by Ref. [33] reported that sorghum silages from ratoon had lower acetic acid (26 g/kg DM) when compared with sorghum silages from the main harvest (41 g/kg DM). The authors found that sorghum plants were less productive in ratoon, besides having lower nutrient values and higher lignin concentration. The same likely happened in the present experiment, where the lower nutrient availability in the sorghum plants during ratoon may have favored the growth of strains better adapted to the environmental conditions (homofermentative strains in this case).

Acetic acid is essential for the AS of the silage as it inhibits the growth of yeasts and filamentous fungi responsible for spoilage when the silage is exposed to air [27,34]. To ensure greater AS, acetic acid levels between 15 and 30 g/kg DM are considered ideal [27]. Levels below 15 g/kg DM may not provide sufficient protection, whereas levels above 30 g/kg DM must be monitored as they may compromise silage palatability. This explains why, overall, the silages of the tested hybrids had low AS values (between 20 and 37 h). Nevertheless, the highest AS values were observed for the silages with the highest acetic acid values (silages produced in Dourados).

In order to attempt to solve the low AS issue in sorghum silages during ratoon, the use of heterofermentative inoculants, such as *Letilactobacillus buchneri*, would be recommended. That microorganism favors acetic acid production, thus extending silage AS and reducing losses after opening [10].

Under ideal conditions, butyric acid concentrations should remain below 5 g/kg DM, whereas values above 10 g/kg DM indicate inadequate fermentation [27]. Similarly, NH_3-N levels below 10 g/kg TN reflect good protein preservation, while values exceeding this threshold suggest excessive proteolysis and reduced silage quality [27]. In the present study, both parameters remained within the recommended ranges, confirming the efficiency of the fermentation process. Among the hybrids evaluated, CMSXS 5044 stood out with a low butyric acid concentration (0.43 g/kg DM)—slightly higher than the average reported by Ref. [29] for the first cut, yet still consistent with high-quality silage standards. These results highlight the strong ensiling potential of biomass sorghum ratoon, reinforcing its value as a strategic forage alternative in systems aiming for multiple harvests within a single growing season.

Hybrids such as BRS 716 and CMSXS5044, which had the highest CP values, are advantageous because protein is a crucial component in ruminant diets, especially during the growth phase and milk production. The CP content is directly related to silage nutritional value as quality protein is essential for good animal performance [35]. On the other hand, hybrids CMSXS7102 and CMSXS7103 had CP values about 7.25% lower than the highest values found, which may limit their use in diets that require greater protein supply.

In contrast, hybrid CMSXS5039 had the highest NFC and starch values, which makes it interesting as those components are highly fermentable by ruminants and, therefore, may provide more energy than the other hybrids. Moreover, CMSXS5039 had the lowest NDF values among all hybrids tested, which suggests greater digestibility and, consequently, greater dietary efficiency of ruminants.

Therefore, from a nutritional standpoint, the choice of sorghum cultivar for ensiling will depend on the production system and the nutritional needs of the herd. Hybrid CMSXS5039, with greater digestibility and energetic value, is more indicated to systems that seek high dietary efficiency. Hybrids such as BRS 716 and CMSXS5044, with higher CP values, are more appropriate for diets that require greater protein supply. Those variations highlight the importance of selecting the hybrid according to the specific requirements of production and local growth conditions [11].

5. Conclusions

Biomass sorghum hybrids demonstrated high dry matter yield potential even under low-rainfall conditions. All evaluated hybrids produced silages with satisfactory fermentation quality, with reduced aerobic stability being the only limitation, which could be mitigated by the use of heterofermentative inoculants. In terms of chemical composition, CMSXS5039 stood out as the most suitable hybrid for diets targeting higher energy supply, given its superior starch and non-fiber carbohydrate contents, while BRS 716 and CMSXS5044 proved more appropriate for diets demanding higher protein levels due to their elevated crude protein concentrations.

Author Contributions: Conceptualization, Y.A.d.S., M.R., G.C. and M.A.P.O.J.; methodology, M.R. and G.C.; validation, A.C.A.O. and T.F.; formal analysis, I.P.d.O.A.; investigation, Y.A.d.S. and M.R.; resources, M.A.P.O.J. and G.C.; data curation, A.C.A.O.; writing—original draft, Y.A.d.S., I.P.d.O.A., and G.R.P.M.; writing—review and editing, M.A.P.O.J., M.R., G.C. and T.F. All authors have read and agreed to the published version of the manuscript.

Funding: This research was funded by Coordination for the Improvement of Higher Education Personnel (CAPES; Brasília, DF, Brazil–n°0001) and the Development of Education, Science and Technology (FUNDECT; Mato Grosso do Sul, MS, Brazil TO: 118/2024 SIAFIC: 813).

Institutional Review Board Statement: Not applicable.

Informed Consent Statement: Not applicable.

Data Availability Statement: The original contributions presented in this study are included in the article. Further inquiries can be directed to the corresponding author.

Acknowledgments: We gratefully acknowledge the support provided by Embrapa Agropecuária Oeste—CPAO, the Federal University of Grande Dourados (UFGD), and the SISPEC network (Network of Smart and Sustainable Livestock Systems, funded by CYTED, ref. 125RT0167), which were essential for the development and execution of this work. We also sincerely thank Latina Seeds for the generous donation of seeds used in the experimental trials.

Conflicts of Interest: Embrapa is a non-profit public institution under the Ministry of Agriculture and Livestock of Brazil. Its work focuses on the development of technologies and knowledge for agriculture and livestock, with an emphasis on sustainability and food security. Dr. Marciana Retore and Dr. Gessí Ceccon are affiliated with Embrapa, which operates on a non-profit basis. Therefore, the authors state that the research was conducted without any commercial or financial relationships that could be construed as a potential conflict of interest.

References

1. Orrico Junior, M.A.P.; Retore, M.; Manarelli, D.M.; De Souza, F.B.; Ledesma, L.L.M.; Orrico, A.C.A. Forage Potential and Silage Quality of Four Varieties of Saccharine Sorghum. *Pesqui. Agropecu. Bras.* **2015**, *50*, 1201–1207. [CrossRef]
2. Castro, M.; Bruzi, A.T.; Nunes, J.A.R.; Parrella, R.A.C.; Lombardi, G.M.R.; Albuquerque, C.J.; Lopes, M. Agronomic and Energetic Potential of Biomass Sorghum Genotypes. *Am. J. Plant Sci.* **2015**, *6*, 1862–1873. [CrossRef]
3. De Queiroz, F.E.; Rocha, V.R.; Monção, F.P.; Rigueira, J.P.S.; Parrella, R.A.d.C.; Rufino, L.D.d.A.; Santos, A.S.d.; Cordeiro, M.W.S. Effect of Row Spacing and Maturity at Harvest on the Fermentative Profile, Aerobic Stability, and Nutritional Characteristics of Biomass Sorghum (BRS 716) Silage in the Semiarid Region of Brazil. *Rev. Bras. Zootec.* **2021**, *50*, e20200254. [CrossRef]

4. Manarelli, D.M.; Orrico, M.A.P.; Retore, M.; de Vargas, F.M.; da Silva, M.S.J.; Orrico, A.C.A.; Borquis, R.R.A.; Crone, C.; de Oliveira Neves, F. Productive Performance and Quantitative Carcass Traits of Lambs Fed Saccharine Sorghum Silage. *Pesqui. Agropecu. Bras.* **2019**, *54*, e00577. [CrossRef]
5. Ra, K.; Shiotsu, F.; Abe, J.; Morita, S. Biomass Yield and Nitrogen Use Efficiency of Cellulosic Energy Crops for Ethanol Production. *Biomass Bioenergy* **2012**, *37*, 330–334. [CrossRef]
6. Souza, Í.D.A.d.S.; Rocha Júnior, V.R.; Monção, F.P.; Parrella, R.A.d.C.; Cordeiro, M.W.S.; Carvalho, C.d.C.S.; Rigueira, J.P.S.; Aspiazú, I.; Silva, F.V.; Delvaux Júnior, N.d.A. Yield and Nutritional Value of Biomass Sorghum (BRS 716) Managed in Different Row Spacing and Maturity at Harvest in the Semiarid Region of Brazil. *Semin. Agrar.* **2021**, *42*, 3463–3484. [CrossRef]
7. Peerzada, A.M.; Ali, H.H.; Chauhan, B.S. Weed Management in Sorghum [*Sorghum bicolor* (L.) Moench] Using Crop Competition: A Review. *Crop Prot.* **2017**, *95*, 74–80. [CrossRef]
8. Lestari, E.; Dewi, I.S.; Yunita, R.; Nur, A. Evaluation of Ratooning Ability in Several Sweet Sorghum (*Sorghum bicolor* [L.] Moench) Mutant Lines. *AIP Conf. Proc.* **2022**, *2462*, 050003. [CrossRef]
9. Zhou, Y.; Huang, J.; Li, Z.; Wu, Y.; Zhang, J.; Zhang, Y. Yield and Quality in Main and Ratoon Crops of Grain Sorghum under Different Nitrogen Rates and Planting Densities. *Front. Plant Sci.* **2022**, *12*, 778663. [CrossRef]
10. Kung, L., Jr.; Shaver, R.D.; Grant, R.J.; Schmidt, R.J. Silage Review: Interpretation of Chemical, Microbial, and Organoleptic Components of Silages. *J. Dairy Sci.* **2018**, *101*, 4020–4033. [CrossRef] [PubMed]
11. Rosa, M.A.B.; Tardin, F.D.; Souza, J.M.d.S.; Santos, J.A.P.; Macedo, T.d.F.; Santos, J.; de Freitas, M.H.; Todescatto, F.; Parrella, R.A.d.C.; Figueiredo, J.E.F.; et al. Characterization of Forage, Sweet and Biomass Sorghum for Agronomic Performance and Ensilability. *Rev. Bras. Milho e Sorgo* **2022**, *21*, 1–22. [CrossRef]
12. Santos, H.G.; Jacomine, P.K.; Anjos, L.H.C.; Oliveira, V.A.; Lumbreras, J.F.; Coelho, M.R.; Almeida, J.A.; Araújo Filho, J.C.; Oliveira, J.B.; Cunha, T.J.F. *Sistema Brasileiro de Classificação de Solos*, 5th ed.; Embrapa: Brasilia, Brazil, 2018; p. 356.
13. Soil Survey. *Keys to Soil Taxonomy*, 12th ed.; USDA—Natural Resources Conservation Service: Washington, DC, USA, 2014.
14. Fietz, C.R.; Fisch, G.F. O Clima Da Região de Dourados, MS. *Embrapa Agropecuária Oeste* **2008**, *92*, 32.
15. Silva, F.C. *Manual de Análises Químicas de Solos, Plantas e Fertilizantes*, 2nd ed.; Silva, F.C., Ed.; Embrapa Informação Tecnológica: Brasilia, Brazil, 2009; ISBN 978-85-7383-430-7.
16. Borém, A.; Pimentel, L.; Parrella, R. *Sorgo: Do Plantio a Colheita*, 1st ed.; UFV: Vicosa, Brazil, 2014.
17. Jobim, C.C.; Nussio, L.G.; Reis, R.A.; Schmidt, P. Avanços Metodológicos Na Avaliação Da Qualidade Da Forragem Conservada Methodological Advances in Evaluation. *Rev. Bras. Zootec.* **2007**, *36*, 101–119. [CrossRef]
18. Playne, M.J.; Mc Donald, P.T. The Buffering Constituents of Herbage and of Silage. *J. Sci. Food Agric.* **1966**, *17*, 264–268. [CrossRef]
19. Pryce, J.D. A Modification of Barker-Summerson Method for the Determination of Lactic Acid. *Analyst* **1969**, *94*, 1151–1152. [CrossRef] [PubMed]
20. Von Pinho, R.G.V.; Vasconcelos, R.C. *Cultura Do Sorgo*, 1st ed.; UFLA: Lavras, Brazil, 2002.
21. Ferraz-Almeida, R.; Albuquerque, C.J.B.; Camargo, R.; Lemes, E.M.; Soares de Faria, R.; Quintão Lana, R.M. Sorghum–Grass Intercropping Systems under Varying Planting Densities in a Semi-Arid Region: Focusing on Soil Carbon and Grain Yield in the Conservation Systems. *Agriculture* **2022**, *12*, 1762. [CrossRef]
22. Oliveira, I.C.M.; Guilhen, J.H.S.; Ribeiro, P.C.d.O.; Gezan, S.A.; Schaffert, R.E.; Simeone, M.L.F.; Damasceno, C.M.B.; Carneiro, J.E.d.S.; Carneiro, P.C.S.; Parrella, R.A.d.C.; et al. Genotype-by-Environment Interaction and Yield Stability Analysis of Biomass Sorghum Hybrids Using Factor Analytic Models and Environmental Covariates. *Field Crops Res.* **2020**, *257*, 107929. [CrossRef]
23. Li, Y.; Li, C.; Cheng, Q.; Duan, F.; Zhai, W.; Li, Z.; Mao, B.; Ding, F.; Kuang, X.; Chen, Z. Estimating Maize Crop Height and Aboveground Biomass Using Multi-Source Unmanned Aerial Vehicle Remote Sensing and Optuna-Optimized Ensemble Learning Algorithms. *Remote Sens.* **2024**, *16*, 3176. [CrossRef]
24. Shah, A.N.; Tanveer, M.; Abbas, A.; Yildirim, M.; Shah, A.A.; Ahmad, M.I.; Wang, Z.; Sun, W.; Song, Y. Combating Dual Challenges in Maize Under High Planting Density: Stem Lodging and Kernel Abortion. *Front. Plant Sci.* **2021**, *12*, 1–19. [CrossRef]
25. Bibi, A.; Sadaqat, H.A.; Tahir, M.H.N.; Fatima Usman, B.; Ali, M. Genetic Analysis of Forage Quality Traits in Sorghum-Sudangrass Hybrids under Water Stress. *J. Anim. Plant Sci.* **2012**, *22*, 1092–1100.
26. Gebrehanna, M.M.; Gordon, R.J.; Madani, A.; VanderZaag, A.C.; Wood, J.D. Silage Effluent Management: A Review. *J. Environ. Manag.* **2014**, *143*, 113–122. [CrossRef] [PubMed]
27. Borreani, G.; Tabacco, E.; Schmidt, R.J.; Holmes, B.J.; Muck, R.E. Silage Review: Factors Affecting Dry Matter and Quality Losses in Silages. *J. Dairy Sci.* **2018**, *101*, 3952–3979. [CrossRef]
28. Kung, L.; Williams, P.; Schmidt, R.J.; Hu, W. A Blend of Essential Plant Oils Used as an Additive to Alter Silage Fermentation or Used as a Feed Additive for Lactating Dairy Cows. *J. Dairy Sci.* **2008**, *91*, 4793–4800. [CrossRef] [PubMed]
29. Silva, Y.A.; Orrico Junior, M.A.P.; Retore, M.; Ceccon, G.; Fernandes, T.; Orrico, A.C.A. Chemical Composition, Fermentation Parameters and Losses of Silages from Different Hybrids of Biomass Sorghum. *Grass Forage Sci.* **2025**, *80*, e12706. [CrossRef]

30. Alves, J.P.; Galeano Jara, E.S.; Orrico Junior, M.A.P.; Fernandes, T.; Retore, M.; Silva, M.S.J.; Orrico, A.C.A.; Lopes, L.S. Th Influence of Plant Age and Microbes-Enzymatic Additives on Fermentation of Total Mixed Ration Silages of Capiaçu Gras (Pennisetum Purpureum, Schum). *Trop. Anim. Sci. J.* **2022**, *45*, 56–63. [CrossRef]
31. Muck, R.E. Silage Microbiology and Its Control through Additives. *Rev. Bras. Zootec.* **2010**, *39*, 183–191. [CrossRef]
32. Orrico Junior, M.A.P.; Vendramini, J.M.B.; Erickson, J.; Moriel, P.; Silveira, M.L.A.; Aguiar, A.D.; Sanchez, J.M.D.; Silva, W.L.; Silv H.M. Nutritive Value and Fermentation Characteristics of Silages Produced from Different Sweet Sorghum Plant Component with or without Microbial Inoculation. *Appl. Anim. Sci.* **2020**, *36*, 777–783. [CrossRef]
33. Mccormick, M.E.; Morris, D.R.; Ackerson, B.A.; Blouin, D.C. Ratoon Cropping Forage Sorghum for Silage: Yield, Fermentatior and Nutrition. *Agron. J.* **1995**, *87*, 952–957. [CrossRef]
34. Wilkinson, J.M.; Rinne, M. Highlights of Progress in Silage Conservation and Future Perspectives. *Grass Forage Sci.* **2018**, *7* 40–52. [CrossRef]
35. NRC. *Nutrient Requirements of Dairy Cattle*, 8th ed.; National Academy Press: Washington, DC, USA, 2021.

fermentation

MDPI

Article

Effect of Harvesting Time on Starch Degradation in Rumen of Whole-Plant Corn and Its Silage

Long Zhang [1,2,3], Shiqin Liu [1,2,3], Xuepeng Wang [1,2,3], He Wang [1,2,3], Songze Li [1,2,3], Yuguo Zhen [1,2,3,4,5,*] and Xuefeng Zhang [1,2,3,4,5,*]

1 Department of Animal Nutrition and Feed Science, College of Animal Science and Technology, Jilin Agricultural University, Changchun 130000, China; zl319174023@163.com (L.Z.); liushiqin0330@163.com (S.L.); uullzz@proton.me (X.W.); wjiajia9993@163.com (H.W.); ze961672411@163.com (S.L.)

2 Jilin Provincial Key Lab of Animal Nutrition and Feed Science, Jilin Agricultural University, Changchun 130000, China

3 Key Laboratory of Animal Production, Product Quality and Security, Ministry of Education, Jilin Agricultural University, Changchun 130000, China

4 JLAU-Borui Dairy Science and Technology R&D Center, Jilin Agricultural University, Changchun 130000, China

5 Postdoctoral Scientific Research Workstation, Feed Engineering Technology Research Center of Jilin Province, Changchun Borui Science & Technology Co., Ltd., Changchun 130000, China

* Correspondence: nickzhen@263.net (Y.Z.); xuefengz@jlau.edu.cn (X.Z.)

Abstract: Whole-plant corn silage is a critical feedstuff in global ruminant production, and its nutrient composition is closely tied to harvest timing. As starch acts as the primary energy source in silage-based diets, investigating changes in starch degradation rate provides a theoretical basis for optimizing the efficient utilization of whole-plant corn and its silage in ruminant production. In this study, whole-plant corn (harvested from the milk stage to full ripening stage) and its corresponding silage were used as experimental materials. An in vitro simulated rumen fermentation system was employed to determine the contents of starch, prolamin, amylose, and amylopectin in the samples. The results showed that with delayed harvest time, starch content in both whole-plant corn and its silage increased significantly; prolamin and amylose contents first decreased, then increased; amylopectin content first rose significantly before decreasing; and both starch disappearance rate and speed exhibited a trend of first increasing, then decreasing. After silage fermentation, the silage had significant increases in starch, amylose, and amylopectin contents, and starch disappearance rate; prolamin content decreased; and starch disappearance speed increased extremely significantly. This study indicates that whole-plant corn harvest time and silage fermentation regulate the ruminal starch degradation pattern by altering starch structure, prolamin content, and the proportion of rapidly degradable starch.

Keywords: whole-plant corn silage; starch structure; rumen fermentation; rumen dynamic degradation model; gliadin

1. Introduction

With the rapid increase in the number of livestock and poultry globally, the problem of competition for grain between humans and livestock is becoming increasingly serious, and animal husbandry is also being affected. The development of animal husbandry is limited by the supply of high-quality silage [1]. Silage technology can greatly alleviate shortages of coarse feed resources in the breeding process [2]. It is therefore widely used and able to preserve green feed well [3]. Whole-plant corn silage is widely used in ruminant production because it has a high nutritional value and good animal palatability [4]. High-quality whole-plant corn silage can not only improve the health of ruminants but also improve the economic benefits of breeding [5].

The main nutritional value of corn silage comes from starch, and its starch utilization efficiency is affected by the structure of the starch [6]. The starch structure of corn silage is affected by the harvesting time of raw materials and the fermentation effect of silage, which further affects the degradation rate of starch in the rumen [7]. During corn growth, starch accumulates continuously and its structure optimizes steadily over time [8]. From the milk stage to the dough stage, starch synthesis and metabolism are highly active: the proportion of amylopectin increases gradually, and starch granules arrange more tightly—both of which enhance the stability of the starch structure [9]. Silage fermentation also exerts a crucial effect. In an anaerobic environment, microorganisms such as lactic acid bacteria proliferate extensively, and the enzymes they secrete can modify starch structure (e.g., hydrolyzing partial glycosidic bonds and altering the amylose-to-amylopectin ratio), thereby influencing the ruminal starch degradation rate [10]. At the same time, the harvest time also affects the starch structure and gliadin content of corn, thus also affecting its rumen disappearance rate [11]. As harvest timing is delayed, the prolamin content in corn kernels increases. Prolamin binds tightly to starch granules, forming starch–protein complexes—analogous to a "protective layer" surrounding starch granules. This binding impedes the access of rumen microorganisms to starch and their subsequent degradation of it, thereby reducing the ruminal starch disappearance rate [12]. At present, the research on starch structure is limited to the influence of internal nutrition structure, unrelated to in vitro fermentation. Therefore, it is of great significance to explore the influencing factors of the starch disappearance rate of whole-plant corn and whole-plant corn silage to make effective use of whole-plant corn and whole-plant corn silage in livestock production.

We hypothesize that the starch structure of whole-plant corn is altered by harvest timing and silage fermentation, and that this modified starch structure can directly influence the ruminal starch degradation rate. In this study, the starch content, gliadin content, amylopectin content, disappearance rate, and rumen dynamic degradation parameters of whole-plant corn silage collected at different harvest times were studied using the wet chemistry method and simulated rumen fermentation system in vitro, which provided a theoretical basis for the effective utilization of whole-plant corn silage in livestock production.

2. Materials and Methods

2.1. Experiment Design

A 2 × 5 full-factor design was adopted in this experiment. Factor 1 was the type of whole-plant corn (raw material and silage), and factor 2 was the harvest time. Silage corn (variety Liangyu 99) was planted in Changchun City, Jilin Province, China, in 2023. Based on the maturity of whole-plant corn, the experiment was performed on 1 September (milk stage), 8 September (early dough stage), 15 September (mid-dough stage), 20 September (late dough stage), and 25 September (full ripening stage) at a stubble height of 40 cm. For each harvest, 20 plants were randomly selected through multipoint random sampling. After cutting took place using the whole-plant corn cutter (Changchun, China) and whole-plant corn was crushed to about 2 cm in length, multipoint sampling was carried out, and then random sampling was performed using the quartering method. Half of the samples were reserved as raw materials, and the other half were sprayed with Lactobacillus plantarum (Borui Science and Technology Co., Ltd., Changchun, China) and then fermented in a 1L PVC small tank for 60 days, and the compaction density was 800 kg/m^3. After fermentation, the silage samples were mixed for the subsequent in vitro simulated rumen fermentation test. Samples were taken at 0, 3, 6, 12, 24, 48, and 72 h after culture.

For each harvest time of whole-plant corn and whole-plant corn silage, three repetitions were set at different in vitro culture time points for subsequent disappearance rate determination.

2.2. Test Animals and Diet Formulation

Rumen fluid was given to three male small-tailed Han sheep (40 ± 2.5 kg) with permanent rumen fistulae. All of the experimental animals were quarantined, fed in a single cage, and allowed to drink freely. The collection of rumen fluid is carried out before morning feeding. The diet formula is shown in Table 1. All of the animal test procedures were performed according to the Guidelines for the Care and Use of Experimental Animals of Jilin Agricultural University (JLAU-ACUC2023-012). This test refers to the China Meat Sheep Feeding Standard (NY/T816-2004) [13], and the formula is shown in Table 1.

Table 1. Dietary formula and nutritional composition (%DM).

Ingredients	Nutritional Compositions %	Ingredients	Nutritional Compositions %
Corn	35.00	Crude protein	14.27
Soybean meal	13.00	Neutral detergent fiber	28.17
Corn germ meal	6.00	Acid detergent fiber	18.35
Corn bran	10.00	Starch	25.87
Peanut hull	23.00	Ether extract	3.94
Bentonite	4.00		
Sucrose	4.00		
[a] Premix	5.00		
Amount	100.00		

[a] Premix composition (per kilogram): Ca^{2+} 110 mg, Cu^{2+} 140 mg, Zn^{2+} 930 mg, Mn^{2+} 600 mg, Co^{2+} 13 mg, I^- 20 mg, Se^{4+} 13 mg, vitamin A 340 KIU, vitamin D_3 120 KIU, and vitamin E 1700 IU.

2.3. In Vitro Rumen Fermentation

Rumen fluid was collected evenly from the rumen of the fistulated sheep using a rigid PVC tube before morning feeding on the day of the experiment. The collected rumen fluid was filtered through four layers of filter mesh into a thermos (preheated to 39 °C) for later use. The buffer solution was prepared in advance following the method described by Menke (1979) [14], and CO_2; was bubbled through it until a color change occurred, after which it was set aside for use. The sampling site was in close proximity to the laboratory, and the experiment was carried out immediately after sampling.

We evenly mixed the filtered rumen fluid and buffer solution in a ratio of 1:2 and continuously introduced CO_2 for 1 min. Then, we added 120 mL of mixed fermentation solution into each in vitro culture flask and continuously introduced CO_2 for 30 s, immediately closed the bottle cap, and then put it into a constant-temperature air bath incubator (Shanghai Hetian Scientific Instruments Co., Ltd. for cultivation, Shanghai, China) (39 °C, 80 rpm).

2.4. Determination Index and Method

2.4.1. Nutrient Index Content, Starch Disappearance Rate, and Disappearance Rate

Crude protein (CP) was determined using the Kelvin method [15]. The in vitro dry matter disappearance rate (IVDMD) and in vitro starch disappearance rate (IVStarchD) were calculated using the formulas outlined by Huang [16] and DeFeo [17], respectively.

2.4.2. Dynamic Degradation Model of Rumen

Ørskov [18] proposed the rumen dynamic degradation model of feed:

$$\mathrm{Deg}\ (t) = a + b(1 - e^{-ct}) \quad (1)$$

Among these parameters, Deg (t) represents the degradation rate of rumen dry matter or specific nutrients in feed samples after *t* hours of incubation in the rumen. This indicator directly reflects the extent to which feed is decomposed and utilized by rumen microorganisms at a specific incubation time point, and serves as a core observed value for evaluating the degradation characteristics of feed. In this formula, a represents the fast degradation fraction (%). Its determination is typically achieved through correction using the "0-hour degradation rate": After placing the feed sample in a rumen nylon bag, the bag is immediately retrieved and processed following the same rinsing and drying procedures as applied to other incubation time points. The measured dry matter loss rate from this process represents the a value. b represents the slow degradation fraction (%). It is calculated as "the potential maximum degradation rate (a + b) minus the rapidly degradable fraction (a)". The potential maximum degradation rate is derived from model fitting after extending the incubation time (typically 72–96 h, until the degradation rate no longer changes significantly). c represents the degradation rate of b (%/h). This parameter is obtained by fitting the measured degradation rates at different incubation time points (e.g., 3, 6, 12, 24, 48, and 72 h; the specific time points should be adjusted based on feed type and pre-experiment results) through nonlinear regression analysis. t represents the incubation time of feed in the rumen (h).

The model accurately describes the disappearance rate of feed ingredients at various time points and is widely used.

2.4.3. Gliadin, Amylopectin, and Amylose Content

The content of relevant indicators in the sample was determined using an immunosorbent assay (Enzyme-linked Biotech Co., Ltd., Shanghai, China); we followed the assay's instructions for use. The core principle of the determination is the "double-antibody sandwich method", which is based on the specific binding reaction between antigen and antibody, as well as the enzyme-catalyzed color reaction.

2.5. *Data Statistics and Analysis*

The relationship between the two factors was analyzed using general linear models in IBM SPSS Statistics 23 software. A statistical analysis model ($Yijk = \mu + Fi + Vj + Fi \times Vj + eijk$) was used to analyze different harvesting times and silage fermentation and their interaction. Duncan's multiple comparison method was used to test the significance of factors 1 and 2 and their interaction (factor 1 × factor 2). $p < 0.05$ was considered significant, and $p < 0.01$ was considered extremely significant.

3. Results

3.1. *Starch and Structural Nutrient Content of Whole-Plant Corn and Whole-Plant Corn Silage*

As can be seen in Table 2, when the harvest time was delayed, the dry matter content and the protein and starch content of whole-plant corn and whole-plant corn silage significantly increased and then decreased ($p < 0.01$), and the starch content of whole-plant corn silage after undergoing silage fermentation was significantly higher than that of whole-plant corn ($p < 0.01$). The gliadin content first decreased and then increased. After silage fermentation, the gliadin content of whole-plant corn silage was significantly lower than that of whole-plant corn on 15 September, 20 September, and 25 September ($p < 0.01$). The amylopectin content first increased and then decreased significantly ($p < 0.01$). The amylopectin content in whole-plant corn silage after undergoing silage fermentation was significantly higher than that in whole-plant corn ($p < 0.01$). The amylose content first decreased and then increased significantly ($p < 0.01$). The amylose content in whole-plant corn silage was significantly higher than that in whole-plant corn silage ($p < 0.01$). Additionally, the contents of various nutrients were significantly affected by the interaction between silage fermentation and harvesting time ($p < 0.01$).

Table 2. Nutrient content of whole-plant corn and whole-plant corn silage (%DM).

	Whole-Plant Corn					Whole-Plant Corn Silage						*p*-Value		
Items Time	**9.1**	**9.8**	**9.15**	**9.20**	**9.25**	**9.1**	**9.8**	**9.15**	**9.20**	**9.25**	**SEM**	**Fermentation**	**Time**	**F × T**
DM (%)	33.59 ^{a}	34.41 ^{ab}	35.13 ^{bc}	36.10 ^{c}	35.04 ^{bc}	30.79 ^{c}	33.89 ^{b}	35.09 ^{ab}	35.69 ^{a}	35.11 ^{ab}	0.28	<0.01	<0.01	<0.01
CP (%)	7.05 ^{c}	7.75 ^{d}	8.16 ^{e}	6.24 ^{b}	6.19 ^{a}	7.64	7.89	8.30	7.79	7.55	0.14	<0.01	<0.01	<0.01
Sta (%)	30.36 ^{d}	30.83 ^{b}	31.98 ^{a}	30.50 ^{c}	30.11 ^{e}	30.13 ^{e}	34.94 ^{b}	36.15 ^{a}	34.48 ^{c}	33.81 ^{d}	0.41	<0.01	<0.01	<0.01
Gli (ng/mL)	467.53 ^{a}	440.76 ^{c}	431.34 ^{bc}	446.76 ^{b}	475.52 ^{a}	466.26 ^{a}	438.23 ^{b}	400.03 ^{d}	417.45 ^{c}	427.48 ^{b}	4.29	<0.01	<0.01	<0.01
AP (mg/g)	21.35 ^{d}	22.29 ^{b}	23.85 ^{a}	22.51 ^{b}	21.81 ^{c}	21.84 ^{d}	25.46 ^{b}	27.43 ^{a}	25.81 ^{b}	24.81 ^{c}	0.37	<0.01	<0.01	<0.01
AS (mg/g)	9.00 ^{a}	8.54 ^{c}	8.11 ^{e}	7.98 ^{d}	8.29 ^{b}	8.27 ^{d}	9.46 ^{a}	8.71 ^{c}	8.66 ^{c}	8.99 ^{b}	0.08	<0.01	<0.01	<0.01

Note: a, b, c, d, and e indicate significant differences in the nutrients between different harvest times under the same treatment, and if the upper corner marks are different, they indicate significant differences ($p < 0.05$).

3.2. In Vitro Dry Matter Disappearance Rate of Whole-Plant Corn and Whole-Plant Corn Silage

As shown in Figure 1, the dry matter disappearance rate of whole-plant corn and whole-plant corn silage first increased and then decreased along with harvesting time. The dry matter disappearance rate of whole-plant corn harvested on 20 September was significantly higher than that harvested at 3 h, 6 h, 12 h, and 24 h ($p < 0.01$), and the dry matter disappearance rate of whole-plant corn silage harvested on 20 September was significantly higher than that harvested at 3 h and 6 h ($p < 0.01$). The dry matter disappearance rate of whole-plant corn silage after undergoing silage fermentation was significantly higher than that of whole-plant corn ($p < 0.01$). Moreover, the dry matter disappearance rate in vitro was significantly affected by the interaction between silage fermentation and harvesting time ($p < 0.05$).

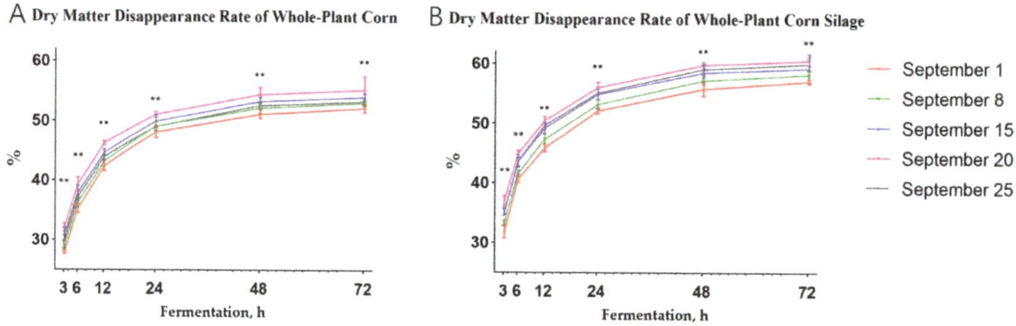

Figure 1. Dry matter disappearance rate of whole-plant corn and whole-plant corn silage (%). (**A**) Dry matter disappearance rate of whole-plant corn; (**B**) dry matter disappearance rate of whole-plant corn silage. Note: ** indicates an extremely significant difference ($p < 0.01$).

3.3. In Vitro Starch Disappearance Rate of Whole-Plant Corn and Whole-Plant Corn Silage

As shown in Figure 2, the starch disappearance rate of whole-plant corn and whole-plant corn silage first increased and then decreased significantly along with the delay in the harvest time ($p < 0.01$). The starch disappearance rate of whole-plant corn and whole-plant corn silage tended to be stable after 24 h. After silage fermentation, the starch disappearance rate of whole-plant corn silage was significantly higher than that of whole-plant corn ($p < 0.01$). Furthermore, the starch disappearance rate in vitro was significantly affected by the interaction between silage fermentation and harvest time ($p < 0.01$).

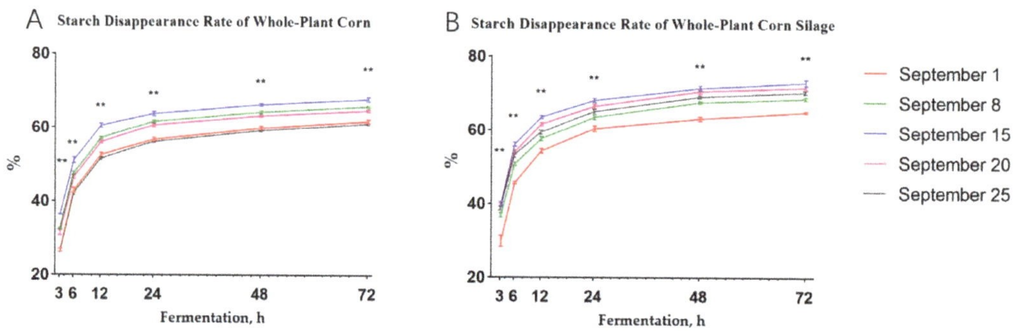

Figure 2. Starch disappearance rate of whole-plant corn and whole-plant corn silage (%). (**A**) Starch disappearance rate of whole-plant corn; (**B**) starch disappearance rate of whole-plant corn silage. Note: ** indicates an extremely significant difference ($p < 0.01$).

3.4. Starch Disappearance Speed of Whole-Plant Corn and Whole-Plant Corn Silage

As shown in Figure 3, the starch disappearance speed of whole-plant corn and whole-plant corn silage first increased and then decreased significantly along with the delay in harvest time ($p < 0.01$). The starch disappearance speed of whole-plant corn and whole-plant corn silage tended to be stable after 24 h. After silage fermentation, the starch disappearance speed of whole-plant corn silage was significantly higher than that of whole-plant corn ($p < 0.01$). The starch disappearance speed was significantly affected by the interaction between silage fermentation and harvest time ($p < 0.01$).

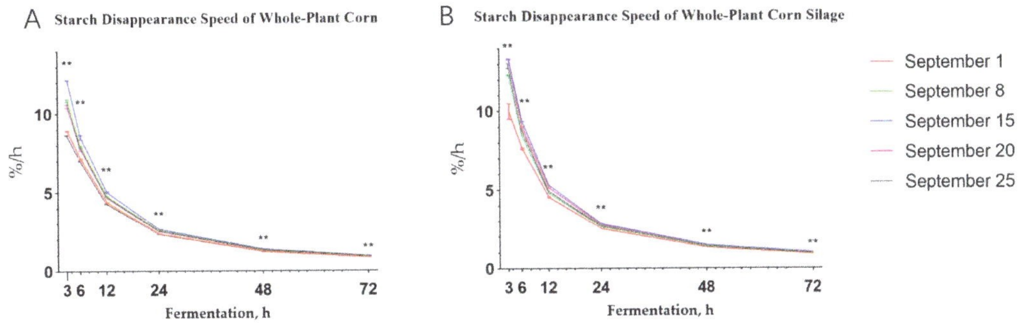

Figure 3. Starch disappearance speed of whole-plant corn and whole-plant corn silage (%/h). (**A**) Starch disappearance speed of whole-plant corn; (**B**) starch disappearance speed of whole-plant corn silage. Note: ** indicates an extremely significant difference ($p < 0.01$).

3.5. Dynamic Degradation Parameters of Whole-Plant Corn and Whole-Plant Corn Silage Starch in Rumen

It can be seen from Table 3 that with the delay in harvesting time, the starch rapid degradation parameter a of whole-plant corn shows a trend of increasing first and then decreasing, and the starch rapid degradation parameter a of whole-plant corn silage generally shows a trend of increasing. The starch slow degradation parameter b of whole-plant corn shows a trend of first decreasing and then increasing, and the starch slow degradation parameter b of whole-plant corn silage generally shows a trend of decreasing. The starch degradation parameter c of whole-plant maize had no obvious change trend. After silage fermentation, the fast degradation parameter a of starch in whole-plant corn silage increased, while the slow degradation parameter b and degradation parameter c decreased.

Table 3. Dynamic degradation parameters of starch in whole-plant corn and whole-plant corn silage in rumen.

	Whole-Plant Corn						Whole-Plant Corn Silage					
Items Time	9.1	9.8	9.15	9.20	9.25	$\mu \pm \sigma$	9.1	9.8	9.15	9.20	9.25	$\mu \pm \sigma$
A (%)	2.93	10.67	12.68	9.58	55.84	7.80 ± 4.01	10.49	23.53	18.09	24.01	24.84	20.19 ± 5.4
B (%)	56.77	53.49	53.46	53.51	55.84	54.61 ± 1.41	52.68	43.68	52.79	46.03	43.83	47.8 ± 4.11
C (%/h)	0.19	0.18	0.20	0.18	0.18	0.19 ± 0.01	0.16	0.14	0.19	0.15	0.14	0.16 ± 0.02

3.6. Loss Rate of Gliadin in Whole-Plant Corn and Whole-Plant Corn Silage

As shown in Figure 4, the gliadin disappearance rate of whole-plant corn and whole-plant corn silage did not change significantly with the delay in harvesting time ($p > 0.05$), but they generally showed a trend of first decreasing and then increasing. There was no significant difference in the gliadin disappearance rate between whole-plant corn and whole-plant corn silage ($p > 0.05$).

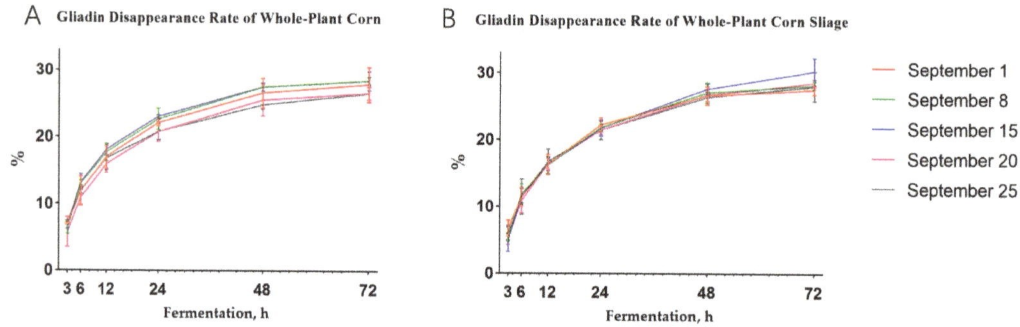

Figure 4. Gliadin disappearance rate of whole-plant corn and whole-plant corn silage (%). (**A**) Gliadin disappearance rate of whole-plant corn; (**B**) gliadin disappearance rate of whole-plant corn silage.

3.7. Loss Rate of Amylose and Amylopectin in Whole-Plant Corn and Whole-Plant Corn Silage

As can be seen from Figure 5, the amylose disappearance rate and amylopectin disappearance rate of whole-plant corn and whole-plant corn silage showed a significant trend of first increasing and then decreasing with the delay in harvesting time ($p < 0.01$). The disappearance rate of amylose and amylopectin in whole-plant corn silage was significantly higher than that in whole-plant corn ($p < 0.01$). The disappearance rate was significantly affected by the interaction between silage fermentation and harvesting time ($p < 0.01$).

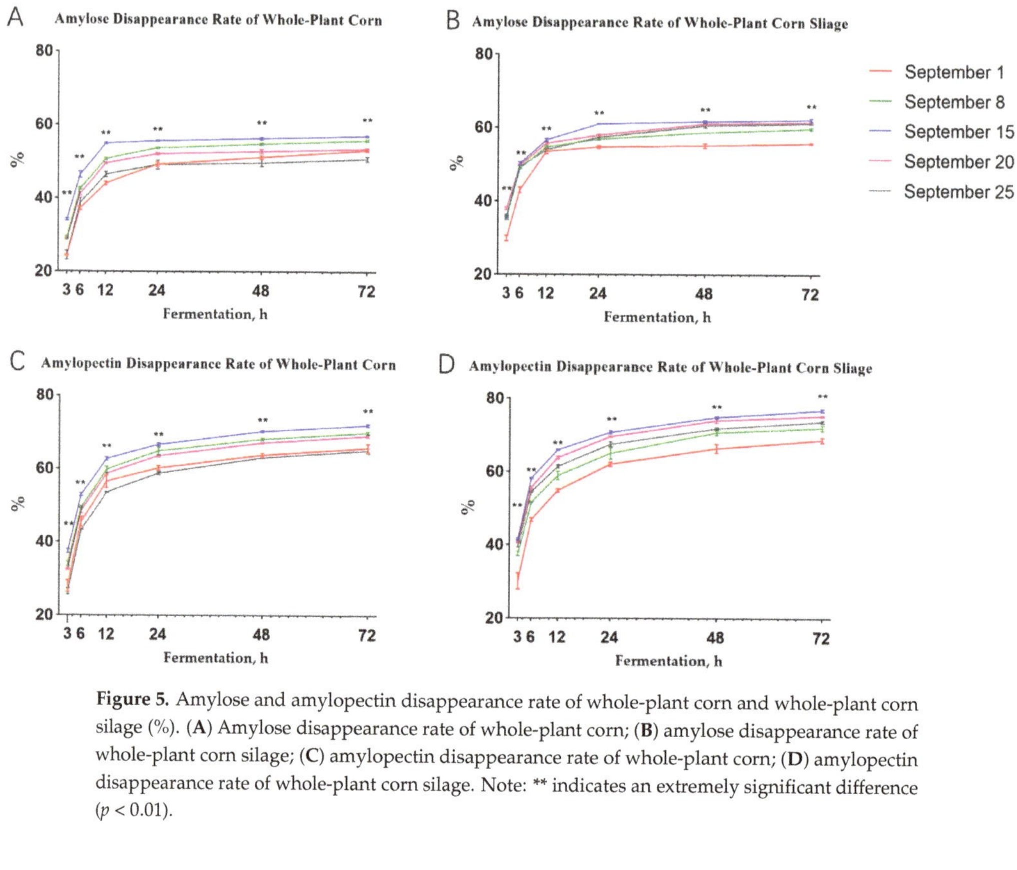

Figure 5. Amylose and amylopectin disappearance rate of whole-plant corn and whole-plant corn silage (%). (**A**) Amylose disappearance rate of whole-plant corn; (**B**) amylose disappearance rate of whole-plant corn silage; (**C**) amylopectin disappearance rate of whole-plant corn; (**D**) amylopectin disappearance rate of whole-plant corn silage. Note: ** indicates an extremely significant difference ($p < 0.01$).

3.8. Loss Speed of Amylose and Amylopectin in Whole-Plant Corn and Whole-Plant Corn Silage

As shown in Figure 6, the disappearance speed of amylose and amylopectin in whole-plant corn and whole-plant corn silage first increased and then decreased significantly with the delay in harvesting time ($p < 0.01$). The disappearance speed of amylose and amylopectin in whole-plant corn silage was significantly higher than that in whole-plant corn ($p < 0.01$). The disappearance speed was significantly affected by the interaction between silage fermentation and harvesting time ($p < 0.01$).

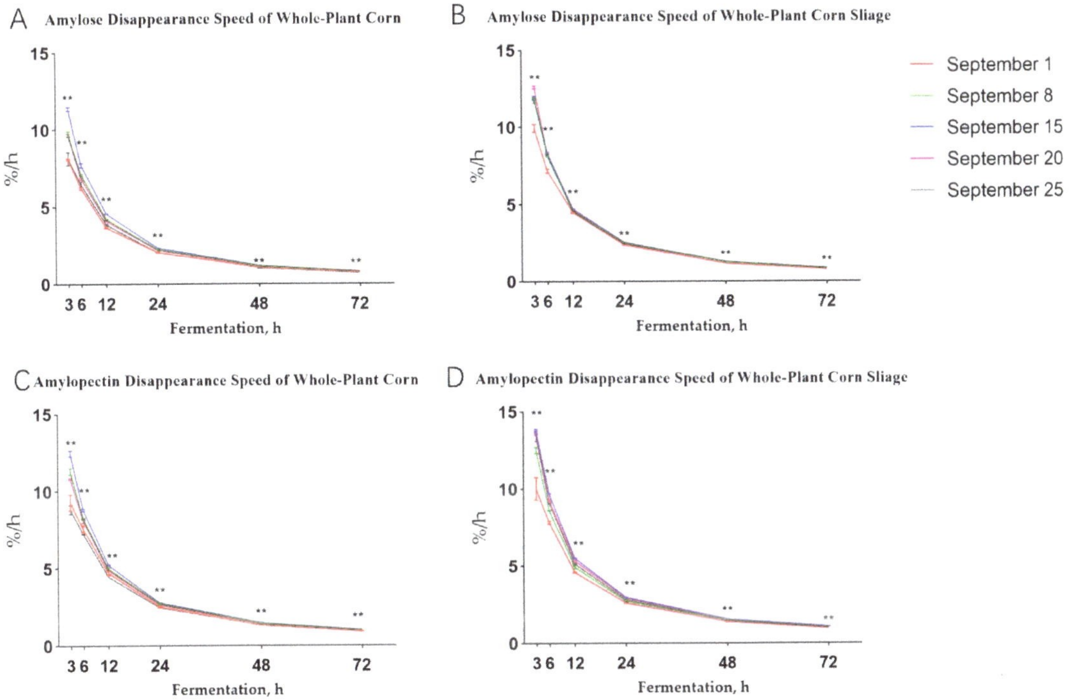

Figure 6. Amylose and amylopectin disappearance speed of whole-plant corn and whole-plant corn silage (%/h). (**A**) Amylose disappearance speed of whole-plant corn; (**B**) amylose disappearance speed of whole-plant corn silage; (**C**) amylopectin disappearance speed of whole-plant corn; (**D**) amylopectin disappearance speed of whole-plant corn silage. Note: ** indicates an extremely significant difference ($p < 0.01$).

3.9. Degradation Parameters of Amylose and Amylopectin in Rumen of Whole-Plant Corn and Whole-Plant Corn Silage

As can be seen in Table 4, with the delay in harvesting time, the amylose rapid degradation parameter a of whole-plant corn generally shows a downward trend, while the rapid degradation parameter a of whole-plant corn silage generally shows an increasing trend. The slow degradation parameter b and degradation parameter c of whole-plant corn generally show an increasing trend, and the slow degradation parameter b and degradation parameter c of whole-plant corn silage generally show a downward trend. After silage fermentation, the rapid degradation parameter a of whole-plant corn silage increased, the slow degradation parameter b decreased, and the degradation parameter c decreased, but this change trend was small. The amylopectin degradation parameter a of whole-plant corn first increased and then decreased, and the rapid degradation parameter a of whole-plant corn silage showed an overall increase. The slow degradation rate parameter b of whole-plant corn first decreased and then increased, the slow degradation rate parameter b of whole-plant corn silage decreased, and the degradation parameter c of whole-plant corn and whole-plant corn silage decreased. After silage fermentation, the amylose degradation parameter a increased, while the amylose degradation parameter b and amylose degradation parameter c decreased. The fast degradation parameter a of amylopectin increased, the slow degradation parameter b decreased, and the degradation parameter c decreased.

Table 4. Dynamic degradation parameters of amylose and amylopectin in whole-plant corn and whole-plant corn silage in rumen.

		Whole-Plant Corn						Whole-Plant Corn Silage					
Items	**Time**	**9.1**	**9.8**	**9.15**	**9.20**	**9.25**	$\mu \pm \sigma$	**9.1**	**9.8**	**9.15**	**9.20**	**9.25**	$\mu \pm \sigma$
Amylose	a (%)	8.90	4.96	6.09	5.95	0.00	5.18 ± 2.90	1.19	13.89	8.80	19.57	16.93	12.08 ± 6.52
	b (%)	42.40	49.70	50.32	46.96	55.00	48.88 ± 4.14	54.66	45.57	51.98	40.41	43.30	47.19 ± 5.34
	c (%/h)	0.16	0.23	0.27	0.23	0.26	0.23 ± 0.04	0.25	0.22	0.25	0.21	0.20	0.22 ± 0.02
Amylopectin	a (%)	0.29	13.00	15.36	10.94	5.32	8.98 ± 5.47	14.14	26.27	20.99	25.05	26.97	22.68 ± 4.75
	b (%)	63.02	54.93	54.39	55.85	57.39	57.12 ± 3.12	52.26	44.11	53.29	48.44	44.80	48.58 ± 3.74
	c (%/h)	0.20	0.17	0.18	0.17	0.16	0.18 ± 0.01	0.14	0.12	0.17	0.14	0.13	0.14 ± 0.02

3.10. Correlation Analysis

It can be seen in Figure 7 that the starch content and starch structure in whole-plant corn and whole-plant corn silage are positively correlated with the rapid starch degradation parameter a, but the gliadin content is negatively correlated with the rapid starch degradation parameter a. Among them, the starch content and amylopectin content have positive correlation coefficients of 0.85 and 0.83, but the gliadin content has a negative correlation coefficient of −0.76. The starch content and starch structure in whole-plant corn and whole-plant corn silage showed a negative correlation with the starch slow degradation parameter b, the gliadin content showed a positive correlation with the starch slow degradation parameter b, and the gliadin content showed a weak positive correlation. Among them, the starch content and amylopectin content had strong negative correlation coefficients of −0.68 and −0.63, and the gliadin content had a weak positive correlation coefficient of 0.50. The degradation parameter c was negatively correlated with the starch content, amylose content, and amylopectin content. The amylose content had a negative correlation coefficient of −0.62 with the degradation parameter c.

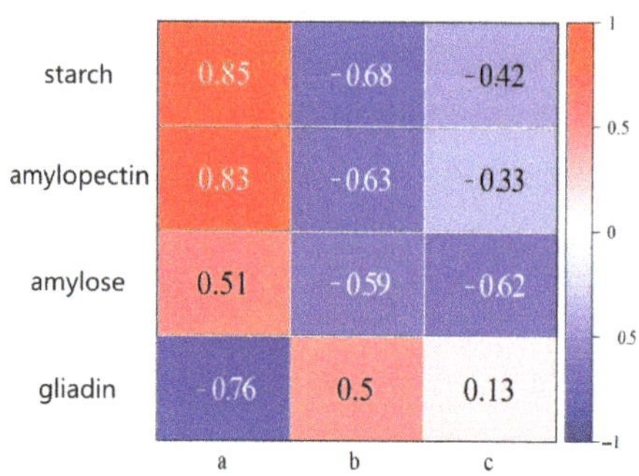

Figure 7. Correlation analysis of starch content, structure, and gliadin content of whole-plant corn and whole-plant corn silage with rumen degradation parameters.

4. Discussion

The harvest time of corn affects its maturity, which directly affects its nutrient content [19]. When harvest time is delayed, corn seeds mature, starch deposition in the seeds continues, and the starch content increases [20]. In this experiment, the starch content of whole-plant corn and whole-plant corn silage reached the highest value on 15 September and then gradually decreased. When the whole-plant corn reached the mature stage, the fiber content gradually increased, and the starch content was also lower than that in the wax mature stage [21]. The whole-plant corn gradually reached the mature stage from 20 September to 25 September, so its starch content was lower than that on 15 September. Harvesting at the right time is key to ensuring a high utilization rate of silage corn [22]. In this experiment, the starch disappearance rate of whole-plant corn and whole-plant corn silage first increased and then decreased with the delay in harvesting time, and the starch disappearance rate of whole-plant corn and whole-plant corn silage reached the highest values on 15 September; these findings are the same as those of Bal, M.A. [23] et al. With the delay in harvest time, the starch content in whole-plant corn and whole-plant corn silage increased, which promoted the synthesis of microbial synthetic protein in the rumen, resulting in an increase in the rumen disappearance rate of starch [24]. With the delay in harvest time, the parameters a and c of starch rapid degradation first increased and then decreased, while the parameter b of starch slow degradation increased gradually. The results showed that the proportion of easily degradable components was higher at the initial stage of maize nutrient accumulation, and with the further increase in maturity, the change in the maize tissue structure led to an increase in degradation resistance, an increase in the proportion of slow degradable components, and a decrease in the proportion of fast degradable components. With the accumulation of the nutrient contents of whole-plant corn and whole-plant corn silage, the slow degradation components decreases, but with the further accumulation of nutrients, it reaches a certain ripening degree, the structural characteristics of the starch change, degradation is inhibited, and its slow degradation components begin to gradually rise [25].

Gliadin can bind to starch to form a protective film, reducing its digestion rate [26]. The removal of gliadin can result in a porous structure being formed on the surface of starch, which can significantly improve its in vitro fermentation characteristics [27]. In this experiment, the gliadin content of whole-plant corn and whole-plant corn silage first decreased and then increased with the delay in harvest time, and the disappearance rate also first decreased and then increased. In this experiment, the content of gliadin in whole-plant corn and whole-plant corn silage decreased in the early harvest stage, and the inhibition of starch degradation decreased. In the late harvest stage, the content of gliadin increased, the binding amount of gliadin and starch increased, and the starch disappearance rate decreased. The content of gliadin in whole-plant corn is higher than that in whole-plant corn silage, which proves that the silage process will destroy and degrade the gliadin structure and make the starch disappearance rate of whole-plant corn silage higher than that of whole-plant corn. This result is the same as the research results of Xu, H [28] and Dehghan-Banadaky M [29]. The disappearance rate of gliadin in the rumen was inversely proportional to the disappearance rate of starch in the rumen, indicating that gliadin could affect the digestibility of starch in the rumen. In addition, gliadin can improve the gel texture characteristics of starch. Zein can attach to the surface of starch molecules, hinder the aggregation and rearrangement of amylose molecules in starch, and make it difficult for water molecules to bind to starch molecules, thus affecting the functional structure of starch [30]. Amylopectin also exhibits a higher degradation rate in the rumen, a specialized digestive environment. The rumen hosts a large community of microorganisms that secrete various hydrolases, and the branched structure of amylopectin renders it a more susceptible target for degradation by these rumen microorganisms [31]. Amylopectin is more easily digested than amylose and consumed more in rumen digestion. In this experiment, the ratio of amylopectin/amylose decreased; this is the same result as that found by Srichuwong et al. [32]. This proportional change may directly impact the energy supply efficiency of whole-plant corn silage. Specifically, amylopectin exhibits high ruminal degradability; thus, a reduction in its relative proportion decreases the "rapid energy-supplying fraction" in the silage, which could prolong the overall starch degradation cycle in the ruminant rumen [33]. Amylose has a "binding" effect on amylopectin. With the increase in the amylose content, the "binding" effect also increases, meaning that amylopectin cannot be fully stretched [34], resulting in a low disappearance rate of amylopectin in early-stage in vitro fermentation. With the delay in harvest time, the rumen starch disappearance rate showed the same change rule as amylopectin but the opposite change rule to amylose. Amylose is generally interspersed with amylopectin during starch granule formation, preventing it from binding to digestive enzymes and reducing starch degradation [35]. In this experiment, the disappearance rate of amylose and amylopectin first increased and then decreased with the delay in harvest time. The amylose and amylopectin content of starch is the main factor determining the starch properties of plants [36]. The delay in harvest time leads to an increase in maize maturity, which affects its amylose and amylopectin content and then affects the degradation rate.

The starch content and starch structure in whole-plant corn and whole-plant corn silage showed a positive correlation with the starch rapid degradation parameter a, whereas gliadin showed a negative correlation with the starch rapid degradation parameter a. The correlation coefficients between the starch content and amylopectin content and rapid degradation parameter a were 0.85 and 0.83, showing a strong positive correlation, indicating that the amylopectin content accounted for more of the effect of the starch content on the rapid degradation parameter a. Due to the special branch structure of amylopectin, amylopectin is easier to digest and improves the starch digestion rate [37], so it has a greater impact on the rapid degradation parameter a. There was a negative correlation between the starch content and starch structure and the starch slow degradation parameter b and a positive correlation between gliadin and the starch slow degradation parameter b. The starch content and amylopectin content had a strong negative correlation with the slow degradation parameter b, indicating that the higher the starch content and amylopectin content, the lower the slow degradation parameter b, and the higher the gliadin content, the higher the slow degradation parameter b and the lower the degradation rate. The degradation parameter c was negatively correlated with the starch content, amylose content, and amylopectin content. The amylose content had a great negative effect on the degradation parameter c, which indicates that the higher the amylose content, the lower the degradation parameter c. Because amylose had a strong negative correlation with the slow degradation parameter b, this affected the degradation parameter c. The correlation coefficient between gliadin and the degradation parameter c was low, indicating that gliadin only had a small effect on the degradation parameter c. In conclusion, the higher the starch content and amylopectin content in whole-plant corn and whole-plant corn silage, the higher the starch disappearance rate; the higher the amylose content, the lower the starch disappearance rate. Gliadin only had a small effect on the starch disappearance rate.

5. Conclusions

Silage fermentation can increase the amylopectin content in whole-plant corn starch, increase the proportion of rapidly degradable starch, reduce the gliadin content, and increase the rumen disappearance rate and the disappearance rate of whole-plant corn silage starch. The deposition of starch and gliadin in corn starch with different structures changed along with different harvest times, which led to the rumen degradation of whole-plant corn silage starch changing along with harvest time.

Author Contributions: L.Z.: writing—original draft, data curation, software, validation, visualization. S.L. (Shiqin Liu): writing—review and editing, resources. L.Z.: writing—review and editing, investigation. X.W.: writing—review and editing, investigation. H.W.: writing—formal analysis, methodology. S.L. (Songze Li): writing—formal analysis, methodology. Y.Z.: writing—review and editing, investigation, funding acquisition. X.Z.: writing—review and editing, supervision. L.Z.: writing—review and editing, formal analysis, methodology. X.Z.: writing—funding acquisition, supervision. X.Z.: writing—review and editing, formal analysis, methodology. X.Z.: resources. All authors have read and agreed to the published version of the manuscript.

Funding: This study was partially supported by the National Key Research and Development Program of China (2023YFD1300905).

Institutional Review Board Statement: The animal study protocol was approved by the animal care committee of Jilin Agricultural University (protocol code: JLAU-ACUC2023-012 and date of approval: 12 December 2023).

Data Availability Statement: The original contributions presented in this study are included in the article. Further inquiries can be directed to the corresponding authors.

Conflicts of Interest: Authors Yuguo Zhen and Xuefeng Zhang are employed by the company "Changchun Borui Science & Technology Co., Ltd.", but for the purposes of this investigation, there was no financing relationship with the company; therefore, there are no conflicts of interest. The remaining authors declare that the research was conducted in the absence of any commercial or financial relationships that could be construed as a potential conflict of interest.

Abbreviations

The following abbreviations are used in this manuscript:

IVDMD	In vitro dry matter disappearance rate
DM	Dry matter content
CP	Crude protein content
Sta	Starch content
Gli	Gliadin content
AP	Amylopectin content
AS	Amylose content
A	Rapid degradation parameter a
B	Slow degradation parameter b
C	Degradation parameter c

References

1. Lv, X.; Chen, L.; Zhou, C.; Zhang, G.; Xie, J.; Kang, J.; Tan, Z.; Tang, S.; Kong, Z.; Liu, Z.; et al. Application of Different Proportions of Sweet Sorghum Silage as a Substitute for Corn Silage in Dairy Cows. *Food Sci. Nutr.* **2023**, *11*, 3575–3587. [CrossRef]
2. Zhang, Y.; Zhao, X.; Chen, W.; Zhou, Z.; Meng, Q.; Wu, H. Effects of Adding Various Silage Additives to Whole Corn Crops at Ensiling on Performance, Rumen Fermentation, and Serum Physiological Characteristics of Growing-Finishing Cattle. *Animals* **2019**, *9*, 695. [CrossRef] [PubMed]
3. Wilkinson, J.M.; Rinne, M. Highlights of Progress in Silage Conservation and Future Perspectives. *Grass Forage Sci.* **2018**, *73*, 40–52. [CrossRef]
4. Rojas-Bourrillon, A.; Russell, J.R.; Trenkle, A.; McGilliard, A.D. Effects of Rolling on the Composition and Utilization by Growing Steers of Whole-Plant Corn Silage. *J. Anim. Sci.* **1987**, *64*, 303–311. [CrossRef]

5. Liu, Y.; Wang, G.; Wu, H.; Meng, Q.; Khan, M.Z.; Zhou, Z. Effect of Hybrid Type on Fermentation and Nutritional Parameters of Whole Plant Corn Silage. *Animals* **2021**, *11*, 1587. [CrossRef]
6. Johnson, L.; Harrison, J.H.; Hunt, C.; Shinners, K.; Doggett, C.G.; Sapienza, D. Nutritive Value of Corn Silage as Affected by Maturity and Mechanical Processing: A Contemporary Review. *J. Dairy Sci.* **1999**, *82*, 2813–2825. [CrossRef]
7. Peyrat, J.; Nozière, P.; Le Morvan, A.; Férard, A.; Protin, P.V.; Baumont, R. Effects of Ensiling Maize and Sample Conditioning on in Situ Rumen Degradation of Dry Matter, Starch and Fibre. *Anim. Feed Sci. Technol.* **2014**, *196*, 12–21. [CrossRef]
8. Ferraretto, L.F.; Crump, P.M.; Shaver, R.D. Effect of Ensiling Time and Exogenous Protease Addition to Whole-Plant Corn Silage of Various Hybrids, Maturities, and Chop Lengths on Nitrogen Fractions and Ruminal in Vitro Starch Digestibility. *J. Dairy Sci.* **2015**, *98*, 8869–8881. [CrossRef]
9. Hasjim, J.; Srichuwong, S.; Scott, M.P.; Jane, J. Kernel Composition, Starch Structure, and Enzyme Digestibility of Opaque-2 Maize and Quality Protein Maize. *J. Agric. Food Chem.* **2009**, *57*, 2049–2055. [CrossRef]
10. Campling, R.C. The Feeding Value of Maize: A Review. *Ann. Appl. Biol.* **1977**, *87*, 284–290. [CrossRef]
11. de Souza, A.M.; Neumann, M.; Rampim, L.; de Almeida, E.R.; Matchula, A.F.; Cristo, F.B.; Faria, M.V. Effect of Storage Time on the Chemical Composition of Whole and Grainless Corn Plant Silage Harvested at Different Maturity Stages. *Rev. Bras. Zootec* **2022**, *51*, e20200180. [CrossRef]
12. Kljak, K.; Grbeša, D.; Duvnjak, M. The Effect of Ensiling on the Starch Digestibility Rate of Rehydrated Grain Silages in Pigs Depends on the Hardness of the Maize Hybrid. *Agriculture* **2025**, *15*, 783. [CrossRef]
13. *NY/T816-2004*; Chinese Feeding Standard for Mutton Sheep. China Agricultural Press: Beijing, China, 2004.
14. Menke, K.H.; Raab, L.; Salewski, A.; Steingass, H.; Fritz, D.; Schneider, W. The Estimation of the Digestibility and Metabolizable Energy Content of Ruminant Feedingstuffs from the Gas Production When They Are Incubated with Rumen Liquor in Vitro. *J. Agric. Sci.* **1979**, *93*, 217–222. [CrossRef]
15. Kirk, P.L. Kjeldahl Method for Total Nitrogen. Available online: https://pubs.acs.org/doi/pdf/10.1021/ac60038a038 (accessed on 16 July 2025).
16. Huang, Z.; Urriola, P.E.; Shurson, G.C. Use of in Vitro Dry Matter Digestibility and Gas Production to Predict Apparent Total Tract Digestibility of Total Dietary Fiber for Growing Pigs. *J. Anim. Sci.* **2017**, *95*, 5474–5484. [CrossRef] [PubMed]
17. DeFeo, M.E.; Shampoe, K.V.; Carvalho, P.H.V.; Silva, F.A.S.; Felix, T.L. In Vitro and in Situ Techniques Yield Different Estimates of Ruminal Disappearance of Barley. *Transl. Anim. Sci.* **2020**, *4*, 141–148. [CrossRef]
18. Ørskov, E.R. The in Situ Technique for the Estimation of Forage Degradability in Ruminants. In *Forage Evaluation in Ruminant Nutrition*; CABI Books: Wallingford, UK, 2000; pp. 175–188. ISBN 978-0-85199-344-7.
19. Russell, J.R. Influence of Harvest Date on the Nutritive Value and Ensiling Characteristics of Maize Stover. *Anim. Feed Sci. Technol.* **1986**, *14*, 11–27. [CrossRef]
20. Xie, Z.L.; Zhang, T.F.; Chen, X.Z.; Li, G.D.; Zhang, J.G. Effects of Maturity Stages on the Nutritive Composition and Silage Quality of Whole Crop Wheat. *Asian-Australas. J. Anim. Sci.* **2012**, *25*, 1374–1380. [CrossRef]
21. Johnson, L.M.; Harrison, J.H.; Davidson, D.; Mahanna, W.C.; Shinners, K.; Linder, D. Corn Silage Management: Effects of Maturity, Inoculation, and Mechanical Processing on Pack Density and Aerobic Stability. *J. Dairy Sci.* **2002**, *85*, 434–444. [CrossRef]
22. Ferraretto, L.F.; Shaver, R.D.; Luck, B.D. *Silage Review:* Recent Advances and Future Technologies for Whole-Plant and Fractionated Corn Silage Harvesting. *J. Dairy Sci.* **2018**, *101*, 3937–3951. [CrossRef]
23. Bal, M.A.; Shaver, R.D.; Shinners, K.J.; Coors, J.G.; Lauer, J.G.; Straub, R.J.; Koegel, R.G. Stage of Maturity, Processing, and Hybrid Effects on Ruminal in Situ Disappearance of Whole-Plant Corn Silage. *Anim. Feed Sci. Technol.* **2000**, *86*, 83–94. [CrossRef]
24. Cui, Y.; Liu, H.; Gao, Z.; Xu, J.; Liu, B.; Guo, M.; Yang, X.; Niu, J.; Zhu, X.; Ma, S.; et al. Whole-Plant Corn Silage Improves Rumen Fermentation and Growth Performance of Beef Cattle by Altering Rumen Microbiota. *Appl. Microbiol. Biotechnol.* **2022**, *106*, 4187–4198. [CrossRef] [PubMed]
25. Philippeau, C.; Landry, J.; Michalet-Doreau, B. Influence of the Protein Distribution of Maize Endosperm on Ruminal Starch Degradability. *J. Sci. Food Agric.* **2000**, *80*, 404–408. [CrossRef]
26. Seifried, N.; Steingaß, H.; Schipprack, W.; Rodehutscord, M. Variation in Ruminal in Situ Degradation of Crude Protein and Starch from Maize Grains Compared to in Vitro Gas Production Kinetics and Physical and Chemical Characteristics. *Arch. Anim. Nutr.* **2016**, *70*, 333–349. [CrossRef]
27. Zadeike, D.; Gaizauskaite, Z.; Svazas, M.; Gruzauskas, R.; Gruzauskas, V.; Damasius, J.; Juodeikiene, G. Application of Solid-State Fermentation for the Improving of Extruded Corn Dry-Milling By-Products and Their Protein Functional Properties. *Life* **2022**, *12*, 1909. [CrossRef]
28. Xu, H.; Zhou, J.; Liu, X.; Yu, J.; Copeland, L.; Wang, S. Methods for Characterizing the Structure of Starch in Relation to Its Applications: A Comprehensive Review. *Crit. Rev. Food Sci. Nutr.* **2023**, *63*, 4799–4816. [CrossRef] [PubMed]

29. Dehghan-banadaky, M.; Corbett, R.; Oba, M. Effects of Barley Grain Processing on Productivity of Cattle. *Anim. Feed Sci. Technol.* **2007**, *137*, 1–24. [CrossRef]
30. Huang, M.; Li, L.; Cheng, X.; Qiu, R.; Wang, Y.; Zhou, X.; Zong, X. Study on the Influence Mechanism of Endogenous Protein Components on the Characteristics of Distiller's Grains Starch. *Food Chem.* **2025**, *486*, 144689. [CrossRef]
31. Piao, M.; Hu, F.; Kong, F.; Liu, Y.; Wang, S.; Cui, K.; Sun, T.; Diao, Q.; Tu, Y. Effects of Dietary Amylose to Amylopectin Ratio on Growth Performance, Carcass Quality Characteristics and Meat Fatty Acids in Chinese Qinchuan Cattle. *J. Integr. Agric.* **2021**, *20*, 3256–3269. [CrossRef]
32. Srichuwong, S.; Sunarti, T.C.; Mishima, T.; Isono, N.; Hisamatsu, M. Starches from Different Botanical Sources I: Contribution of Amylopectin Fine Structure to Thermal Properties and Enzyme Digestibility. *Carbohydr. Polym.* **2005**, *60*, 529–538. [CrossRef]
33. Zhao, F.; Ren, W.; Zhang, A.; Jiang, N.; Liu, W.; Wang, F. Effects of Different Amylose to Amylopectin Ratios on Rumen Fermentation and Development in Fattening Lambs. *Asian-Australas. J. Anim. Sci.* **2018**, *31*, 1611–1618. [CrossRef]
34. Yan, X.; Wu, Z.-Z.; Li, M.-Y.; Yin, F.; Ren, K.-X.; Tao, H. The Combined Effects of Extrusion and Heat-Moisture Treatment on the Physicochemical Properties and Digestibility of Corn Starch. *Int. J. Biol. Macromol.* **2019**, *134*, 1108–1112. [CrossRef]
35. Stevnebø, A.; Sahlström, S.; Svihus, B. Starch Structure and Degree of Starch Hydrolysis of Small and Large Starch Granules from Barley Varieties with Varying Amylose Content. *Anim. Feed Sci. Technol.* **2006**, *130*, 23–38. [CrossRef]
36. French, D. Chemical and Physical Properties of Starch. *J. Anim. Sci.* **1973**, *37*, 1048–1061. [CrossRef]
37. Zhu, F. Relationships between Amylopectin Internal Molecular Structure and Physicochemical Properties of Starch. *Trends Food Sci. Technol.* **2018**, *78*, 234–242. [CrossRef]

fermentation

MDPI

Article

Assessment of Effects of Storage Time on Fermentation Profile, Chemical Composition, Bacterial Community Structure, Co-Occurrence Network, and Pathogenic Risk in Corn Stover Silage

Zhumei Du [1], Ying Meng [1], Yifan Chen [1], Shaojuan Cui [1], Siran Wang [2,]* and Xuebing Yan [1,]*

[1] College of Animal Science and Technology, Yangzhou University, Yangzhou 225009, China; duzhumei2021@126.com (Z.D.); mx120230940@stu.yzu.edu.cn (Y.M.); mz120241609@stu.yzu.edu.cn (Y.C.); 17862186278@163.com (S.C.)

[2] Institute of Animal Science, Jiangsu Academy of Agricultural Sciences, Nanjing 210014, China

* Correspondence: wangsiran@jaas.ac.cn (S.W.); yxbbjzz@163.com (X.Y.)

Abstract: In order to achieve the efficient utilization of agricultural by-products and overcome the bottleneck of animal feed shortages in dry seasons, this study utilized corn stover (CS; *Zea mays* L.) as a material to systematically investigate the dynamic changes in the fermentation quality, bacterial community structure, and pathogenic risk of silage under different fermentation times (0, 3, 7, 15, and 30 days). CS has high nutritive value, including crude protein and sugar, and can serve as a carbon source and a nitrogen source for silage fermentation. After ensiling, CS silage (CSTS) exhibited excellent fermentation quality, characterized by relatively high lactic acid content, low pH, and ammonia nitrogen content within an acceptable range. In addition, neither propionic acid nor butyric acid was detected in any of the silages. CS exhibited high α-diversity, with *Serratia marcescens* being the dominant bacterial species. After ensiling, the α-diversity significantly ($p < 0.05$) decreased, and *Lactiplantibacillus plantarum* was the dominant species during the fermentation process. With the extension of fermentation days, the relative abundance of *Lactiplantibacillus plantarum* significantly ($p < 0.05$) increased, reaching a peak and stabilizing between 15 and 30 days. Ultimately, lactic acid bacteria dominated and constructed a microbial symbiotic network system. In the bacterial community of CSTS, the abundance of "potential pathogens" was significantly ($p < 0.01$) lower than that of CS. These results provide data support for establishing a microbial regulation theory for silage fermentation, thereby improving the basic research system for the biological conversion of agricultural by-products and alleviating feed shortages in dry seasons.

Keywords: bacterial community structure; corn stover; crop by-product; microbial network system; silage fermentation dynamics

1. Introduction

In tropical developing countries, the primary roughage sources for ruminants are local native grasses and agricultural by-products [1]. Insufficient forage quality and quantity, particularly during the dry season, have become key constraints to the development of livestock production in this region [2]. Forage crops and grass typically grow well in the summer and autumn when water sources are abundant. Their dry matter (DM) yield and nutritive value are at high levels, resulting in a surplus supply and adequate nutrition for cattle and sheep. However, during dry seasons such as winter and spring, a severe shortage of feed supply can cause cattle and sheep to lose weight or even die [3]. Therefore, it is particularly important to develop efficient technologies to optimize the utilization of grasses and crop by-products, thereby effectively ensuring the nutritional requirements for animal growth [4].

Corn (*Zea mays* L.) as an important food and cash crop, is widely cultivated due to its high yield potential and broad adaptability, growing from tropical to temperate regions, making it one of the most widely cultivated crops globally (second only to wheat and rice) [5]. During mechanized corn harvesting, after the kernels are separated from the cobs by the threshing mechanism, the stover is left in the field due to the lack of efficient methods for its utilization [6]. Therefore, converting by-products into usable resources has become a key strategy to alleviate resource constraints and improve the ecological environment. Corn stover (CS) has a high fiber content, and direct feeding with this by-product poses issues such as poor palatability and low digestibility. Using technology for the production and storage of silage is the most effective means to enhance the crude protein (CP) digestibility of straw and reduce DM loss rates [7]. Therefore, developing efficient technologies to optimize the utilization of grasses and crop by-products, thereby effectively ensuring the nutritional requirements for animal growth, is of particular importance [4].

Traditional microbial plate culture methods struggle to fully analyze the dynamic changes in microbial communities within complex silage systems. Although next-generation high-throughput sequencing technology has been widely applied in studies of microbial population structure, its resolution is limited to the genus level. PacBio's single-molecule real-time (SMRT) sequencing technology has the advantage of longer read lengths and enables microbial classification at the species level, significantly enhancing detection sensitivity and accuracy [8]. This study employed a combination of multi-omics technologies and conventional physicochemical analyses to systematically elucidate the interactive mechanisms underlying bacterial community succession, nutrient transformation, and the evolution of physicochemical properties during CS fermentation. A quantitative assessment model linking fermentation duration to silage quality was constructed to provide precise fermentation cycle regulation strategies for the efficient utilization of CS.

2. Materials and Methods

2.1. CS Silage (CSTS) Sample Preparation

Corn (cv. Futian 301) harvesting was carried out at the Fresh Corn Introduction Trial Base (Yangzhou, Jiangsu, China) in 20 October 2023. After harvesting corn grain-bearing ears with a harvester (HF441G, Iseki Co., Ltd., Matsuyama, Tokyo, Japan), the remaining CS left in the field was harvested manually with a sickle. Subsequently, a forage chopper (CX-201, Yamamoto Co., Ltd., Tendo, Japan) was employed to chop the CS residue into lengths of 1–2 cm. After thorough mixing, samples of 1 kg were weighed out in triplicate. The samples were then immediately placed in a 4 °C freezer and transported to the laboratory for an analysis of chemical composition and bacterial community.

This study utilized four CSTS treatments: CSTS-3 (3 days of fermentation), CSTS-7 (7 days), CSTS-15 (15 days), and CSTS-30 (30 days). Polyethylene plastic bags of consistent specifications (FB2025, 350 mm × 500 mm, Fisher Scientific, Hampton, NY, USA) were selected to ensure uniformity of the experimental conditions. By thoroughly mixing the CS, four independent replicates were set up for each of the four CSTS treatments, with each replicate corresponding to one silage bag, resulting in a total of 16 silage bags (4 treatments × 4 replicates). Approximately 1 kg of CS was placed in each silage bag, and then the bags were vacuum-sealed for 1 min using a vacuum-packing sealer machine

(AliceV952S, ASONE, Osaka, Japan). After sealing, the bags were placed in a room-temperature environment of 19–28 °C and fermented for 3, 7, 15, and 30 days, respectively.

After fermentation was complete, three samples were randomly selected from each treatment at the different fermentation periods for opening and sampling. Subsequently, after thorough mixing and homogenization, each treatment sample was divided into 3 portions: the first portion (approximately 50 g) was stored in a −80 °C freezer for use in PacBio SMRT sequencing technology; the second portion (approximately 200 g) was dried for chemical composition analysis; and the remaining portion (approximately 10 g) was used to prepare liquid filtrates for the analysis of microbial population and fermentation quality.

2.2. *Chemical Composition, Fermentation Quality, and Microbial Population*

The CS and CSTS were dried in a 65 °C constant-temperature oven for 48 h to achieve a constant weight state. Subsequently, the dried samples in each bag were thoroughly mixed, and then ground using a mill (T1-200; CMT Co., Ltd., Tokyo, Japan) until all of the samples could pass through a 1 mm sieve for subsequent experimental analysis. The DM, organic matter (OM), CP, ether extract (EE), neutral detergent fiber (NDF), and acid detergent fiber (ADF) were determined by the AOAC method [9]. Water-soluble carbohydrate (WSC) was determined by High-Performance Liquid Chromatography (HPLC, LC-2000 plus, JASCO Corporation, Tokyo, Japan) [10].

Cold-water extraction was used to determine fermentation indicators, including pH, ammonia nitrogen (NH_3-N), and organic acids, in the silages [10]. Briefly, 10 g of sample was homogenized with 90 mL of deionized water and refrigerated at 4 °C for 24 h to prepare the extraction solution. The pH was directly determined by a pH meter (S-90, Mettler Toledo, Zurich, Switzerland). The phenol-sodium hypochlorite colorimetric method was used to measure the NH_3-N content [10]. Organic acids were analyzed via HPLC (Agilent 1260 Infinity II, Santa Clara, CA, USA), with the machine equipped with a Phenomenex ReZex ROA column (7.8 mm × 300 mm; Phenomenex, Torrance, CA, USA). The chromatographic conditions included an oven temperature of 60 °C, UV detection of 450 nm, and isocratic elution with 3 mmol/L $HClO_4$ of 1.0 mL/min.

Samples of 10 g were blended with 90 mL of 0.85% sterilized saline solution. Then, the mixture was shaken for 5 min using a Vortex-Genie Blender (Scientific Industries, Bohemia, New York, NY, USA), and the supernatant was subjected to serial dilutions ranging from 10^{-1} to 10^{-8}. The lactic acid bacteria (LAB) were determined using MRS medium (Difco Laboratories, Detroit, MI, USA); aerobic bacteria using nutrient agar medium (Nissui-Seiyaku Co., Ltd., Tokyo, Japan); coliform bacteria using blue-light broth agar medium (Nissui-Seiyaku Co., Ltd.); and yeast and mold using potato dextrose agar medium (Nissui-Seiyaku Co., Ltd.) [11]. The microbial count was performed using the spread plate method, with three replicates set for each dilution gradient. The results are expressed as lg colony-forming units (cfu)/g of fresh matter (FM).

2.3. *Measurement of Bacterial Community Structure by SMRT*

Sample preparation: A 10 g sample (CS or CSTS) was mixed with 90 mL of 0.85% sterilized saline solution and shaken at 120 rpm for 2 h, with 3 replicates per sample. The mixture was filtered through four layers of pre-sterilized gauze, and the filtrate was centrifuged at 10,000 rpm for 10 min at 4 °C. The total genomic DNA was extracted using the DNA kit (DP302-02, Tiangen kit, Beijing, China) according to the method described by Yan et al. [12].

Library construction: Specific primers containing barcodes were synthesized based on the full-length primer sequence, and PCR technology was used to amplify the extracted DNA [12]. After purification, quantification, and homogenization, the samples were used to construct a SMRTbell sequencing library. Following quality control of the library, it was sequenced using the PacBio Sequel II platform (Pacific Biosciences of California, Inc., Menlo Park, CA, USA).

Bioinformatic analysis: The raw data was stored in BAM format and converted to circular consensus sequence (CCS) files using SMRT Link 8.0 software from Pacific Biosciences [13]. The samples were separated by barcodes, and FASTQ format data was generated for subsequent analysis. Based on the SILVA reference database (version 138.1), operational taxonomic unit (OTU) classification and annotation were conducted by employing the naive Bayes classifier integrated into QIIME2. Venn diagrams were constructed using the R package (version 1.2) to describe the shared and unique microorganisms in all of the samples [14]. Using the QIIME software (version 2020.6.0), the alpha diversity indices, including abundance-based coverage estimator (ACE), Chao1, Shannon, and Simpson, were computed [15]. The relative abundance of the microbial community at the genus and species level were analyzed using the Excel Statistical Package for Windows (version 2003). Microbial network analysis was conducted and hierarchical clustering heatmaps were constructed using PyNAST v1.2.2 and the pvclust package of R v3.0.2 [16,17]. Bugbase phenotypic prediction analysis was performed using the Bugbase tool (version 0.1.0) [18].

2.4. Statistical Analysis

One-way analysis of variance (ANOVA) with the general linear model (GLM) procedure in SPSS (Version 19.0, SPSS Inc., Chicago, IL, USA) was employed to assess variations in chemical composition and fermentation quality across treatments [19]. All data were analyzed using orthogonal contrasts to evaluate linear and quadratic responses to fermentation duration, followed by Tukey's multiple comparison test to assess statistical significance ($p < 0.05$).

3. Results

Table 1 presents the chemical composition and microbial population of CS. DM and OM accounted for over 31% of FM and 91% of DM, respectively. The CP and EE contents in CS were below 7.0% and 2.1% of DM, respectively. The NDF, ADF, and WSC contents exceeded 54%, 28%, and 6.6% of DM, respectively. The microbial counts of CS showed that LAB, coliform bacteria, and yeast exceeded 5.0 lg cfu/g FM; aerobic bacteria exceeded 7.0 lg cfu/g FM; and mold was below 4.0 lg cfu/g FM.

Table 1. Chemical composition and microbial population of CS.

Items	CS
Chemical composition	
DM (%)	31.21 ± 0.22
OM (% DM)	91.38 ± 1.14
CP (% DM)	7.01 ± 0.04
EE (% DM)	2.08 ± 0.03
NDF (% DM)	54.95 ± 1.23
ADF (% DM)	28.74 ± 0.67
WSC (% DM)	6.65 ± 0.30
Microbial population (lg cfu/g FM)	
Lactic acid bacteria	5.84 ± 0.10
Aerobic bacteria	7.54 ± 0.12
Coliform bacteria	5.31 ± 0.19
Yeast	5.05 ± 0.06
Mold	4.22 ± 0.11

Data are expressed as mean ± standard deviation, and statistics are based on three independent replicate samples. ADF, acid detergent fiber; cfu, colony-forming units; CS, corn stover; CP, crude protein; DM, dry matter; EE, ether extract; FM, fresh matter; NDF, neutral detergent fiber; OM, organic matter; WSC, water-soluble carbohydrate.

The chemical composition, fermentation quality, and microbial population for CSTS fermented for 3, 7, 15, and 30 days are shown in Table 2. In terms of chemical composition, with the extension of fermentation duration, CP, EE, NDF, and ADF contents were linearly ($p < 0.05$) decreased, while DM and OM contents were consistently maintained at a stable level, with no significant ($p > 0.05$) differences observed. For fermentation quality, with the extension of fermentation duration, pH value and NH_3-N content showed linear ($p < 0.05$) and quadratic ($p < 0.05$) decreases, whereas lactic acid content showed a linear ($p < 0.05$) and quadratic ($p < 0.05$) increase. Specifically, the lowest pH and NH_3-N content, along with the highest lactic acid content, were observed in the CSTS-15 treatment. With the extension of fermentation duration, acetic acid showed a linear ($p < 0.05$) increase, while propionic acid and butyric acid were not detected during the entire period. For the microbial population, with the extension of fermentation duration, LAB count was linearly ($p < 0.05$) and quadratic ($p < 0.05$) increased, whereas the counts of aerobic bacteria and yeast were linearly ($p < 0.05$) or quadratic ($p < 0.05$) decreased. The highest LAB count and the lowest counts of aerobic bacteria and yeast were found in the CSTS-15 treatment. No coliform bacteria or mold (<2.0 lg cfu/g FM) was enumerated.

Table 2. Chemical composition, fermentation quality, and microbial population of CSTS for different periods of time.

Items	Treatments				SEM	*p* Value for Contrasts	
	CSTS-3	CSTS-7	CSTS-15	CSTS-30		*P*-L	*P*-Q
Chemical composition							
DM (%)	31.82	31.22	31.97	31.49	0.31	0.86	0.86
OM (% DM)	91.05	90.93	90.89	91.00	0.23	0.86	0.63
CP (% DM)	6.17 ^{a}	5.76 ^{ab}	5.66 ^{ab}	5.28 ^{b}	0.18	<0.01	0.96
EE (% DM)	2.02 ^{a}	1.94 ^{ab}	1.78 ^{bc}	1.65 ^{c}	0.05	<0.01	0.69
NDF (% DM)	53.54 ^{a}	52.76 ^{ab}	52.27 ^{b}	52.10 ^{b}	0.24	<0.01	0.24
ADF (% DM)	28.26 ^{a}	27.94 ^{ab}	27.63 ^{bc}	27.31 ^{c}	0.12	<0.01	0.99
Fermentation quality							
pH	3.88 ^{a}	3.80 ^{a}	3.53 ^{c}	3.64 ^{b}	0.02	<0.01	<0.01
Lactic acid (% DM)	0.62 ^{b}	1.84 ^{a}	2.03 ^{a}	1.90 ^{a}	0.07	<0.01	<0.01
Acetic acid (% DM)	0.32 ^{b}	0.57 ^{a}	0.56 ^{a}	0.59 ^{a}	0.05	<0.01	0.41
Propionic acid (% DM)	ND	ND	ND	ND	--	--	--
Butyric acid (% DM)	ND	ND	ND	ND	--	--	--
NH_3-N (% DM)	0.83 ^{a}	0.77 ^{ab}	0.66 ^{c}	0.72 ^{bc}	0.02	<0.01	0.02
Microbial population (lg cfu/g FM)							
Lactic acid bacteria	5.35 ^{c}	6.16 ^{b}	6.51 ^{a}	6.45 ^{a}	0.05	<0.01	<0.01
Aerobic bacteria	6.19 ^{a}	5.68 ^{b}	5.10 ^{c}	5.13 ^{c}	0.05	<0.01	<0.01
Coliform bacteria	ND	ND	ND	ND	--	--	--
Yeast	5.52	5.30	5.24	5.36	0.07	0.13	0.04
Mold	ND	ND	ND	ND	--	--	--

Data are the average of three samples. $^{a–c}$ Values with different superscripts in the same row indicate statistically significant differences among different fermentation days ($p < 0.05$). *P*-L and *P*-Q represent linear and quadratic effects of fermentation day. CSTS, corn stover silage; CSTS-3, corn stover fermented for 3 days; CSTS-7, corn stover fermented for 7 days; CSTS-15, corn stover fermented for 15 days; CSTS-30, corn stover fermented for 30 days. --, value is zero; ADF, acid detergent fiber; cfu, colony-forming units; CP, crude protein; DM, dry matter; EE, ether extract; FM, fresh matter; ND, not detected; NDF, neutral detergent fiber; NH_3-N, ammonia nitrogen; OM, organic matter; SEM, standard error of the mean.

The number of OTUs and their overlapping relationships for CS and CSTS treated with different fermentation durations are shown in Figure 1. The number of OTUs in CS was significantly ($p < 0.05$) higher than that in all of the silages (Figure 1A). In Figure 1B, the number of shared OTUs between CS and CSTS treatments was 148, while the number of unique OTUs in CS was significantly ($p < 0.05$) higher than that in the CSTS treatment.

Figure 1C shows that the four treatments, CSTS-3, CSTS-7, CSTS-15, and CSTS-30, shared 67 OTUs. As the fermentation duration extended, the number of unique OTUs in each treatment generally increased, but the CSTS-15 treatment exhibited the lowest number of unique OTUs.

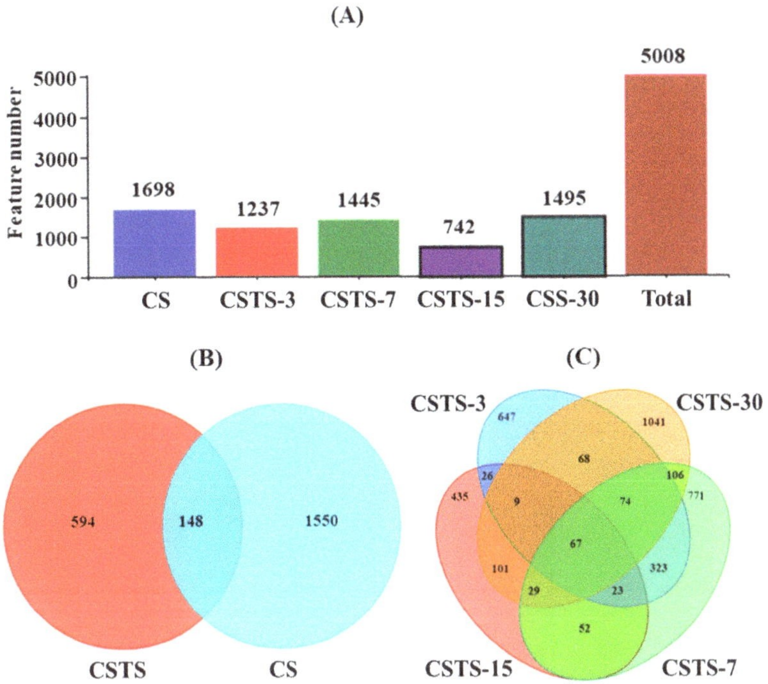

Figure 1. Intergroup features and overlapping relationships of OTU numbers in CS before and after ensiling. (**A**) Feature number of CS, CSTS-3, CSTS-7, CSTS-15, and CSTS-30; (**B**) OTU of CS and CSTS; (**C**) OTU of all CSTS treatments. CS, corn stover; CSTS, corn stover silage; CSTS-3, corn stover fermented for 3 days; CSTS-7, corn stover fermented for 7 days; CSTS-15, corn stover fermented for 15 days; CSTS-30, corn stover fermented for 30 days.

Figure 2 shows the changes in the α-diversity indices of CS before and after ensiling. Compared with the CS treatment, the ACE indices of the CSTS-3, CSTS-7, and CSTS-30 treatments significantly ($p < 0.05$) increased, while there were no marked ($p > 0.05$) differences with the CSTS-15 and CSTS-30 treatments (Figure 2A). There were no significant ($p > 0.05$) differences in the Chao 1 indices between CS and all of the CSTSs (Figure 2B). Compared with the CS, the Simpson and Shannon indices were significantly ($p < 0.05$) lower in all of the silages, with the lowest indices observed in the CSTS-15 treatment (Figure 2C,D).

The relative abundance of the bacterial community at the genus and species levels in the CS and CSTS treatments is shown in Figures 3A and 3B, respectively. Before ensiling, the dominant bacterial species was *Serratia marcescens* (genus *Serratia*). After ensiling, *Lactiplantibacillus plantarum* (genus *Lactiplantibacillus*) became the dominant species in the CSTS-3, CSTS-7, CSTS-15, and CSTS-30 treatments. As the ensiling days increased, the relative abundance of *Lactiplantibacillus plantarum* in CSTS rose steadily from day 3 to day 15 (CSTS-3, CSTS-7, and CSTS-15), and then stabilized at later stages (CSTS-15 and CSTS-30).

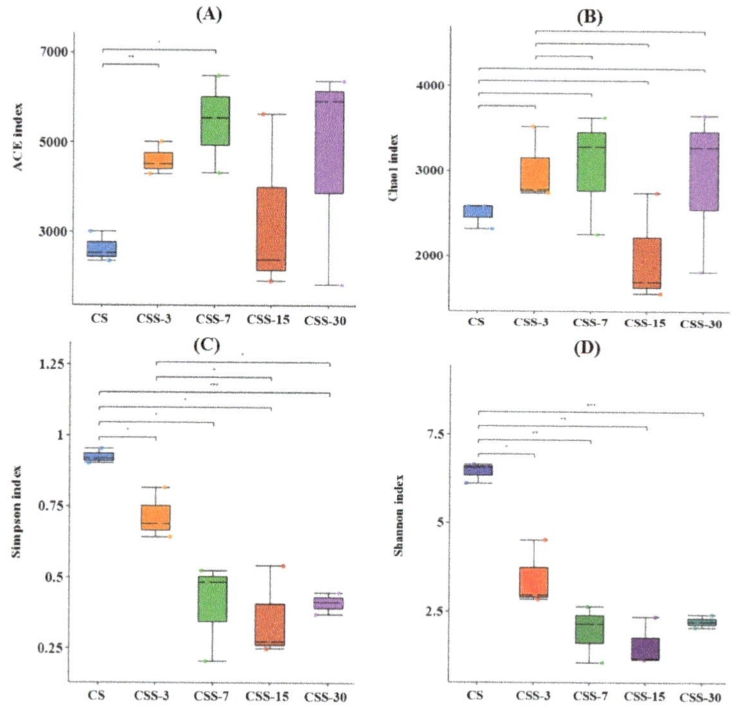

Figure 2. General information on alpha diversity analysis of CS before and after ensiling. (**A**) ACE index of alpha diversity; (**B**) Chao 1 index of alpha diversity; (**C**) Simpson index of alpha diversity; (**D**) Shannon index of alpha diversity. ACE, abundance-based coverage estimator; CS, corn stover; CSTS, corn stover silage; CSTS-3, corn stover fermented for 3 days; CSTS-7, corn stover fermented for 7 days; CSTS-15, corn stover fermented for 15 days; CSTS-30, corn stover fermented for 30 days. *, $p < 0.05$; **, $p < 0.01$; ***, $p < 0.001$.

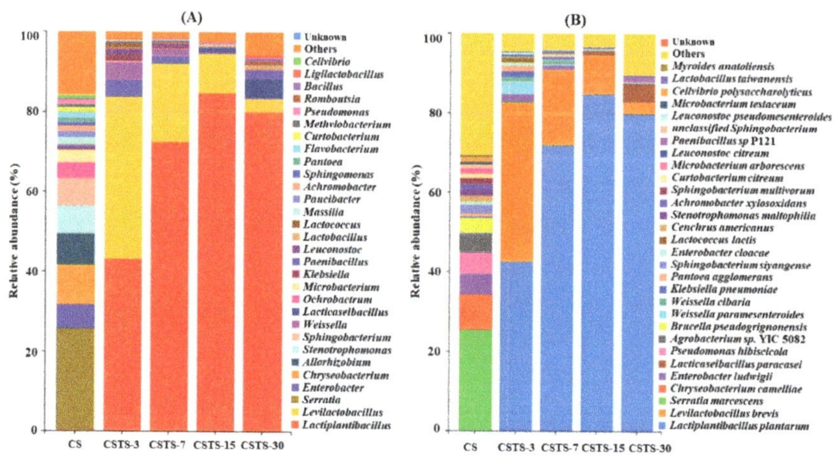

Figure 3. Relative abundance of the top 30 bacterial communities associated with CS at the genus (**A**) and species (**B**) levels before and after ensiling. CS, corn stover; CSTS, corn stover silage; CSTS-3, corn stover fermented for 3 days; CSTS-7, corn stover fermented for 7 days; CSTS-15, corn stover fermented for 15 days; CSTS-30, corn stover fermented for 30 days.

Figure 4A shows the correlation analyses of the network of microbes related to CSTS at the species level. The key species in CSTS was *Lactobacillus spicheri*, which was positively correlated with *Levilactobacillus hammesii*. *Lentilactobacillus kefiri* was negatively correlated with *Clostridium beijerinckii*. Figure 4B displays the correlation heatmap and hierarchical cluster analysis of the bacterial community and terminal fermentation products in CSTS. Lactic acid and LAB were strongly and positively correlated with *Lactiplantibacillus plantarum*, and pH was negatively correlated with *Weissella paramesenteroides*.

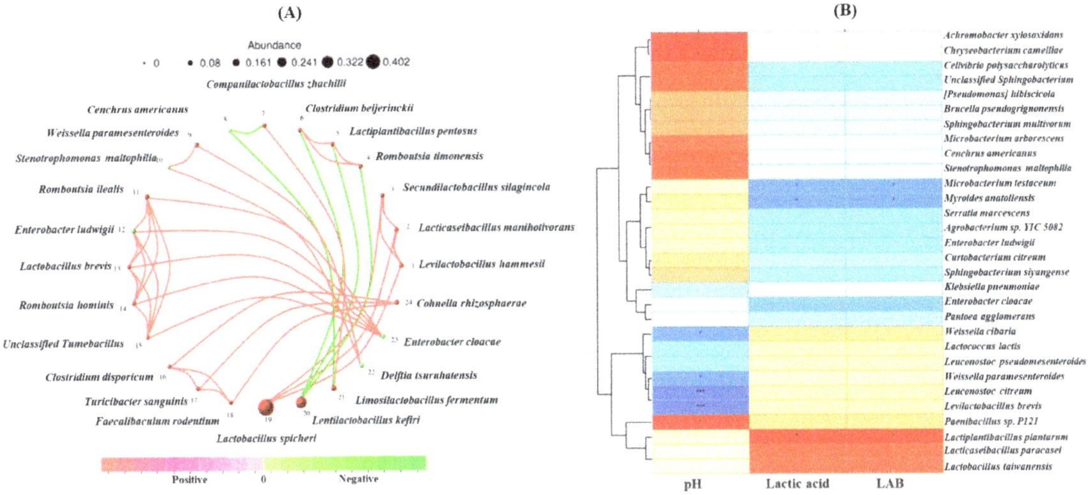

Figure 4. Species-level correlation network (**A**) and correlation analysis (**B**) between bacterial community and terminal fermentation product. LAB, lactic acid bacteria. *, $p < 0.05$, ***, $p < 0.0001$.

Figure 5 shows the results of the Bugbase phenotypic prediction analysis before and after the ensiling of CS. Compared with the CS, the proportion of Gram-positive bacteria in the CSTS treatment significantly ($p < 0.05$) increased, while the proportions of Gram-negative bacteria and potential pathogenic bacteria significantly ($p < 0.05$) decreased.

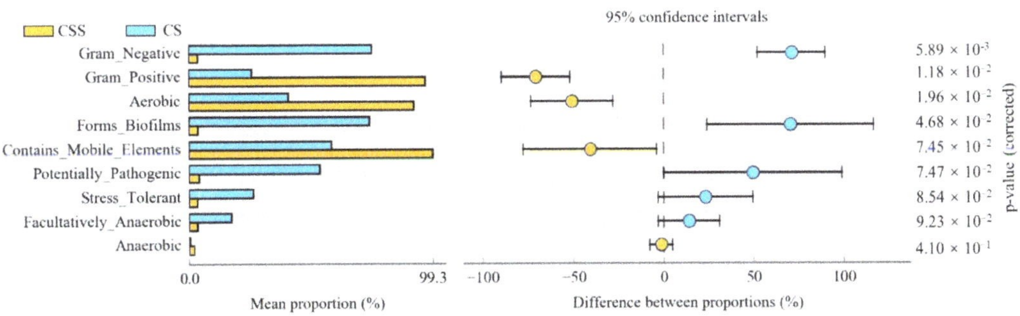

Figure 5. Bugbase phenotype prediction analysis of CS before and after ensiling. CS, corn stover; CSTS, corn stover silage.

4. Discussion

4.1. Biochemical Composition of CS

CS, as the residual part of crops after harvesting the seeds, is a renewable agricultural by-product with a huge yield and rich in a variety of micronutrients [20]. The traditional way of utilizing this by-product is incineration, but the incineration process releases large amounts of CO_2, CO, SO_2, N_2O and other greenhouse gases [21]. The utilization of CSTS can not only effectively reduce the amount of agricultural waste generated, but also achieve the recycling of resources, providing strong support for the sustainable development of green agriculture [7]. Crop by-products are usually dried naturally in the field after harvesting and preserved as hay; however, nutrient losses and degradation can occur. Also, drying artificially is energetically expensive [22]. Therefore, silage fermentation is considered to be the ideal technology for feed harvest and storage. Theoretically, the ideal conditions for silage fermentation include a moisture of 60–70%, WSC content greater than 5% of DM, and a LAB count exceeding 10^5 cfu/g of FM [23]. It was evident from the results of this study that CS had a low moisture content as well as a high WSC content and LAB count, indicating relatively high nutritional value. Therefore, preserving and utilizing the nutrient content of these agricultural by-products with the help of silage fermentation is important to enhance feed quality and livestock productivity.

4.2. Fermentation Quality of Silage

During the fermentation process, LAB ferment sugars to lactic acid, thereby lowering the pH, inhibiting the growth of harmful bacteria, and ultimately producing high-quality silage [24]. During the ensiling of CS in this study, oxygen in the silage environment was continuously consumed, leading to the death of aerobic microorganisms and promoting the growth of LAB. This resulted in a continuous decrease in pH and the aerobic bacteria count, while the lactic acid content and LAB count increased, indicating good fermentation quality. Additionally, after 15 days of fermentation, the silage environment stabilized, resulting in peak fermentation quality.

4.3. Microbial Alpha Diversity and Community Dynamics

The dynamic evolution of bacterial community structure during silage production is characterized by significant changes in microbial abundance and α-diversity. This process is affected by a number of factors, including material composition, environmental parameters, and microbial interactions, directly impacting the fermentation quality of silage [25]. In the field of bacterial community research, traditional plate culture techniques and modern high-throughput sequencing technologies exhibit significant technical differences and functional complementarity. Traditional plate culture techniques are limited by the cultivability of microorganisms, typically only enabling genus-level identification. For some groups, species-level classification requires additional physiological and biochemical tests, making it difficult to fully reflect the true composition of the community [26]. In contrast, modern high-throughput sequencing technologies based on NGS and SMRT have overcome the cultivability bottleneck, enabling the analysis of the entire community structure, including uncultivable microorganisms, at the genus and species levels, with resolution reaching the species or even strain level. Additionally, these technologies can perform multidimensional analyses of community diversity, abundance distribution, and functional gene potential [27].

The number of OTUs combined with the ACE and Chao1 indices is commonly used to assess the species richness of bacterial communities, while the Shannon and Simpson indices are used to measure species diversity [28]. The results of this study showed that the moisture of the CS was approximately 78%, which falls within the high moisture range, providing an ideal environment for the proliferation of aerobic microorganisms (Figures 1 and 2). Under adequate oxygen supply, the bacterial community on the surface of CS exhibited high species richness and diversity. As the silage process progressed, oxygen in the system was rapidly depleted, while acidic substances produced during fermentation continued to accumulate. Under the dual environmental stresses of anaerobic conditions and low pH, Gram-negative bacteria struggle to adapt to environmental changes and undergo significant mortality. By the end of fermentation, acid-tolerant and anaerobic Gram-positive LAB emerge as dominant, establishing a dominant position through vigorous lactic acid fermentation. This dynamic succession of microbial community was characterized by LAB gradually replacing other microbial groups in the CS, becoming the dominant population, ultimately leading to a significant reduction in microbial abundance and diversity in CSTS.

Analyzing the community structure composition and dynamic succession at the genus and species levels is crucial for precisely regulating fermentation quality and optimizing process parameters. Figure 3 of this study shows that *Serratia marcescens* was the initial dominant bacterial species in CS. As a Gram-negative opportunistic pathogen belonging to the Enterobacteriaceae family, it is widely distributed in soil, water bodies, and the surfaces of plants and animals, and commonly causes hospital-acquired infections in critically ill patients or immunocompromised individuals [29,30]. As silage fermentation progressed, *Lactiplantibacillus plantarum* became the dominant bacterial community in CSTS. Additionally, its relative abundance continued to increase with fermentation time. This phenomenon was attributed to the strong acid tolerance of *Lactiplantibacillus plantarum*, which enables it to rapidly respond to the dual stresses of anaerobic and acidic conditions within the silage system. [12]. As fermentation time extended, the pH within the silage environment continued to decrease. *Lactiplantibacillus plantarum* further intensified the acidic environment through vigorous lactic acid fermentation, forming a positive feedback loop of acid production–bacterial inhibition. This not only effectively suppressed the proliferation of harmful bacteria such as *Serratia marcescens*, but also enhanced the nutrient retention and storage stability of silage by maintaining a low-pH steady state [31].

4.4. Correlation Analysis Between Microbial Network and Fermentation Product

Silage fermentation is a dynamic process regulated by the synergistic interaction between microbial communities and metabolic products [32]. During ensiling, various microorganisms utilize distinct metabolic pathways to convert substrates into a wide range of metabolic products, directly determining the fermentation quality of silage. Concurrently, these metabolic products act as environmental regulatory factors, reshaping the structure of microbial communities through mechanisms such as pH changes and antibacterial effects [23]. This bidirectional interaction between microbial metabolism and environmental feedback constitutes the core mechanism of silage fermentation, ultimately determining key quality indicators such as the nutritive value, shelf life, and palatability of silage [28].

Lactobacillus spicheri, as a facultative anaerobic homofermentative LAB, rapidly accumulated lactic acid to decrease the pH of silage, thereby safeguarding feed nutrients. This bacterium often establishes a symbiotic relationship with the heterofermentative LAB *Levilactobacillus hammesii* through a lactic acid–acetic acid feedback metabolic mechanism, thereby optimizing the composition of fermentation products [33,34]. *Lentilactobacillus kefiri*, a Gram-positive heterofermentative LAB, exhibits inhibitory activity against pathogenic bacteria like *Bacillus cereus* and *Staphylococcus aureus*. This is achieved through the production of bioactive components during fermentation, thereby enhancing the safety and quality attributes of fermented food products [35]. During ensiling, the proliferation of

Clostridium beijerinckii can trigger butyric acid-driven spoilage, a process that depletes nutrients and generates undesirable odors [36]. Consequentially, a negative associative pattern was observed between *Lentilactobacillus kefiri* and *Clostridium beijerinckii*, underscoring their opposing roles in shaping silage fermentation outcomes. *Lactiplantibacillus plantarum* in silage can accelerate the rapid fermentation of LAB, produce lactic acid, and lower the pH, thereby inhibiting the reproduction of plant enzymes and harmful bacteria and preserving the nutrients associated with ensiling [37]. *Weissella paramesenteroides* is a heterofermentative LAB, and the acetic acid accumulated during ensiling can effectively improve the aerobic stability of silage. Homo- and hetero-fermented LAB synergistically improve the silage quality [38].

4.5. Bugbase Phenotypes Predicted in the Bacterial Community

By leveraging microbial genome sequence information such as gene functional annotation and metabolic pathways, Bugbase can accurately predict the phenotypic characteristics of unknown strains from multiple perspectives, including cellular structure, metabolic function, ecology, and pathogenicity. This tool has demonstrated its application value in multiple fields. In silage research, its core functions focus on fermentative bacterial community structure analysis, metabolic function prediction, and safety assessment [18].

In this study, damage to CS during harvesting and transportation released juices containing carbon sources, mainly including glucose and amino acids. These substances promoted the proliferation of Gram-negative bacteria inherent in the plant surface. At the same time, due to the presence of pathogenic bacteria in the bacterial community, the CS posed a biosafety risk [39]. However, since the silage system is mainly dominated by obligate anaerobic bacteria and facultative anaerobic bacteria, these bacterial communities efficiently accumulate lactic acid through glycolysis, causing the pH of the system to decrease and forming an acidic microenvironment, thereby inhibiting the growth of acid-sensitive Gram-negative bacteria [40]. This process ultimately promoted Gram-positive bacteria to become the dominant bacterial community and stabilized colonization, thereby improving the fermentation quality and reducing the potential pathogenic risk of CSTS.

5. Conclusions

This study indicated that CS contained carbon and nitrogen sources suitable for natural silage fermentation, with the primary microbial community in the CS being *Serratia marcescens*, which shifted to *Lactiplantibacillus plantarum* as the fermentation progressed. Also, the optimal coupling period for microbial metabolism and nutrient transformation occurred between 15 and 30 days. This study provided a theoretical foundation for improving the research framework for the biological conversion of agricultural by-products.

Author Contributions: Methodology, formal analysis, visualization, and writing—original draft, Z.D.; validation, methodology, and software, Z.D., Y.M. and Y.C.; investigation and formal analysis, S.C.; methodology and visualization, S.W. and X.Y.; conceptualization, project administration, resources, data curation, supervision, funding acquisition, and writing—review, and editing, Z.D., S.W. and X.Y. All authors have read and agreed to the published version of the manuscript.

Funding: This work was supported by the Project of the National Natural Science Foundation of China (grant numbers 32401483) and High-level Talents Program of "Lv-Yang-Jin-Feng of Yangzhou" (grant numbers 137012846/5019).

Institutional Review Board Statement: Not applicable.

Informed Consent Statement: Not applicable.

Data Availability Statement: The original contributions presented in this study are included in the article. Further inquiries can be directed to the corresponding authors.

Conflicts of Interest: The authors declare no conflicts of interest.

References

1. Boval, M.; Dixon, R.M. The importance of grasslands for animal production and other functions: A review on management and methodological progress in the tropics. *Animal* **2012**, *6*, 748–762. [CrossRef] [PubMed]
2. Larsen, R.E.; Shapero, M.W.K.; Striby, K.; Althouse, L.D.; Meade, D.E.; Brown, K.; Horney, M.R.; Rao, D.R.; Davy, J.S.; Rigby, C.W.; et al. Forage quantity and quality dynamics due to weathering over the dry season on California annual rangelands. *Rangel. Ecol. Manag.* **2021**, *76*, 150–156. [CrossRef]
3. Hogan, J.P.; Phillips, C.J.C. Starvation of ruminant livestock. In *Nutrition and the Welfare of Farm Animals. Animal Welfare*; Springer: Cham, Switzerland, 2016.
4. Dunière, L.; Sindou, J.; Chaucheyras-Durand, F.; Chevallier, I.; Thévenot-Sergentet, D. Silage processing and strategies to prevent persistence of undesirable microorganisms. *Anim. Feed Sci. Technol.* **2013**, *182*, 1–15. [CrossRef]
5. Ranum, P.; Peña-Rosas, J.P.; Garcia-Casal, M.N. Global maize production, utilization, and consumption. *Ann. N. Y. Acad. Sci.* **2014**, *1312*, 105–112. [CrossRef] [PubMed]
6. Chen, J.; Stokes, M.R.; Wallace, C.R. Effects of enzyme-inoculant systems on preservation and nutritive value of hay crop and corn silages. *J. Dairy Sci.* **1994**, *77*, 501–512. [CrossRef] [PubMed]
7. Gao, J.L.; Wang, P.; Zhou, C.H.; Li, P.; Tang, H.Y.; Zhang, J.B.; Cai, Y. Chemical composition and in vitro digestibility of corn stover during field exposure and their fermentation characteristics of silage prepared with microbial additives. *Asian-Australas. J. Anim. Sci.* **2019**, *32*, 1854–1863. [CrossRef] [PubMed]
8. Mosher, J.J.; Bowman, B.; Bernberg, E.L.; Shevchenko, O.; Kan, J.; Korlach, J.; Kaplan, L.A. Improved performance of the PacBio SMRT technology for 16S rDNA sequencing. *J. Microbiol. Meth.* **2014**, *104*, 59–60. [CrossRef] [PubMed]
9. AOAC. *Official Methods of Analysis*, 17th ed.; Association of Official Analytical Chemists: Arlington, VA, USA, 2000.
10. Cai, Y. Analysis method for silage. In *Field and Laboratory Methods for Grassland Science*; Tosho Printing Co., Ltd.: Tokyo, Japan, 2004.
11. Cai, Y.; Benno, Y.; Ogawa, M.; Kumai, S. Effect of applying lactic acid bacteria isolated from forage crops on fermentation characteristics and aerobic deterioration of silage. *J. Dairy Sci.* **1999**, *82*, 520–526. [CrossRef] [PubMed]
12. Yan, Y.H.; Li, X.M.; Guan, H.; Huang, L.K.; Ma, X.; Peng, Y.; Li, Z.; Nie, G.; Zhou, J.Q.; Yang, W.Y. Microbial community and fermentation characteristic of Italian ryegrass silage prepared with corn stover and lactic acid bacteria. *Bioresource Technol.* **2019**, *279*, 166–173. [CrossRef] [PubMed]
13. Hou, Q.; Xu, H.; Zheng, Y.; Xi, X.; Kwok, L.Y.; Sun, Z.; Zhang, H.; Zhang, W. Evaluation of bacterial contamination in raw milk, ultra-high temperature milk and infant formula using single molecule, real-time sequencing technology. *J. Dairy Sci.* **2015**, *98*, 8464–8472. [CrossRef] [PubMed]
14. Shade, A.; Handselman, J. Beyond the Venn diagram: The hunt for a core microbiome. *Environ. Microbiol.* **2012**, *14*, 4–12. [CrossRef] [PubMed]
15. Price, M.N.; Dehal, P.S.; Arkin, A.P. FastTree: Computing large minimum evolution trees with profiles instead of a distance matrix. *Mol. Biol. Evol.* **2009**, *26*, 1641–1650. [CrossRef] [PubMed]
16. Langfelder, P.; Horvath, S. WGCNA: An R package for weighted correlation network analysis. *BMC Bioinform.* **2008**, *9*, 559. [CrossRef] [PubMed]
17. Suzuki, R.; Shimodaira, H. Pvclust: An R package for assessing the uncertainty in hierarchical clustering. *Bioinformatics* **2006**, *22*, 1540–1542. [CrossRef] [PubMed]
18. Kennang Ouamba, A.J.; Gagnon, M.; Varin, T.; Chouinard, P.Y.; LaPointe, G.; Roy, D. Metataxonomic insights into the microbial ecology of farm-scale hay, grass or legume, and corn silage produced with and without inoculants. *Front. Syst. Biol.* **2022**, *2*, 955611. [CrossRef]
19. Steel, R.G.; Torrie, J.H. *Principles and Procedures of Statistics: A Biometrical Approach*, 1st ed.; McGraw Hill Company: New York, NY, USA, 1980.
20. Ginni, G.; Kavitha, S.; Yukesh Kannah, R.; Shashi Kant, B.; Adish Kumar, S.; Rajkumar, M.; Gopalakrishnan, K.; Arivalagan, P.; Nguyen Thuy Lan, C.; Rajesh Banu, J. Valorization of agricultural residues: Different biorefinery routes. *J. Environ. Chem. Eng.* **2021**, *9*, 105435. [CrossRef]
21. Filonchyk, M.; Peterson, M.P.; Zhang, L.; Hurynovich, V.; He, Y. Greenhouse gases emissions and global climate change: Examining the influence of CO_2, CH_4, and N_2O. *Sci. Total Environ.* **2024**, *935*, 173359. [CrossRef] [PubMed]

22. Scarbrough, D.A.; Coblentz, W.K.; Coffey, K.P.; Turner, J.E.; Davis, G.V.; Kellogg, D.W.; Hellwig, D.H. Effects of summe management and fall harvest date on ruminal in situ degradation of crude protein in stockpiled bermudagrass. *Anim. Feed Sc Technol.* **2002**, *96*, 119–133. [CrossRef]
23. Kung, L., Jr.; Shaver, R.D.; Grant, R.J.; Schmidt, R.J. Silage review: Interpretation of chemical, microbial, and organolepti components of silages. *J. Dairy Sci.* **2018**, *101*, 4020–4033. [CrossRef] [PubMed]
24. Okoye, C.O.; Wang, Y.; Gao, L.; Wu, Y.; Li, X.; Sun, J.; Jiang, J. The performance of lactic acid bacteria in silage production: review of modern biotechnology for silage improvement. *Microbiol. Res.* **2023**, *266*, 127212. [CrossRef] [PubMed]
25. Ren, F.; He, R.; Zhou, X.; Gu, Q.; Xia, Z.; Liang, M.; Zhou, J.; Lin, B.; Zou, C. Dynamic changes in fermentation profiles an bacterial community composition during sugarcane top silage fermentation: A preliminary study. *Bioresour. Technol.* **2019**, *28* 121315. [CrossRef] [PubMed]
26. McAllister, T.A.; Dunière, L.; Drouin, P.; Xu, S.; Wang, Y.; Munns, K.; Zaheer, R. Silage review: Using molecular approaches t define the microbial ecology of silage. *J. Dairy Sci.* **2018**, *101*, 4060–4074. [CrossRef] [PubMed]
27. Guo, L.; Wang, X.; Lin, Y.; Yang, X.; Ni, K.; Yang, F. Microorganisms that are critical for the fermentation quality of paper mulberr silage. *Food Energy Secur.* **2021**, *10*, e304. [CrossRef]
28. Xu, D.; Wang, N.; Rinne, M.; Ke, W.; Weinberg, Z.G.; Da, M.; Bai, J.; Zhang, Y.; Li, F.; Guo, X. The bacterial community an metabolome dynamics and their interactions modulate fermentation process of whole crop corn silage prepared with or withou inoculants. *Microb. Biotechnol.* **2021**, *14*, 561–576. [CrossRef] [PubMed]
29. Zivkovic Zaric, R.; Zaric, M.; Sekulic, M.; Zornic, N.; Nesic, J.; Rosic, V.; Vulovic, T.; Spasic, M.; Vuleta, M.; Jovanovic, J.; et a Antimicrobial treatment of Serratia marcescens invasive infections: Systematic review. *Antibiotics* **2023**, *12*, 367. [CrossRef [PubMed]
30. Khanna, A.; Khanna, M.; Aggarwal, A. Serratia marcescens-a rare opportunistic nosocomial pathogen and measures to limit it spread in hospitalized patients. *J. Clin. Diagn. Res.* **2013**, *7*, 243–246. [PubMed]
31. Rocchetti, M.T.; Russo, P.; Capozzi, V.; Drider, D.; Spano, G.; Fiocco, D. Bioprospecting antimicrobials from *Lactiplantibacillu plantarum*: Key factors underlying its probiotic action. *Int. J. Mol. Sci.* **2021**, *22*, 12076. [CrossRef] [PubMed]
32. Wang, T.; Zhang, J.; Shi, W.; Sun, J.; Xia, T.; Huang, F.; Liu, Y.; Li, H.; Teng, K.; Zhong, J. Dynamic changes in fermentatior quality and structure and function of the microbiome during mixed silage of *Sesbania cannabina* and sweet sorghum grown o saline-alkaline land. *Microbiol. Spectr.* **2022**, *10*, e0248322. [CrossRef] [PubMed]
33. Ianniello, R.G.; Zheng, J.; Zotta, T.; Ricciardi, A.; Gänzle, M.G. Biochemical analysis of respiratory metabolism in the heterofer mentative *Lactobacillus spicheri* and *Lactobacillus reuteri*. *J. Appl. Microbiol.* **2015**, *119*, 763–775. [CrossRef] [PubMed]
34. Valcheva, R.; Korakli, M.; Onno, B.; Prévost, H.; Ivanova, I.; Ehrmann, M.A.; Dousset, X.; Gänzle, M.G.; Vogel, R.F. *Lactobacillus hammesii* sp. nov. isolated from French sourdough. *Int. J. Syst. Evol. Microbiol.* **2005**, *55 Pt 2*, 763–767. [CrossRef] [PubMed]
35. Carasi, P.; Malamud, M.; Serradell, M.A. Potentiality of food-isolated *Lentilactobacillus kefiri* strains as probiotics: State-of-art anc perspectives. *Curr. Microbiol.* **2022**, *79*, 21. [CrossRef] [PubMed]
36. Mermejo, B.C.; Bortolucci, J.; de Andrade, A.R.; Reginatto, V. The non-solventogenic *Clostridium beijerinckii* Br 21 produces 1,3-propanediol from glycerol with butyrate as the main by-product. *Front. Sustain. Food Syst.* **2022**, *6*, 848022. [CrossRef]
37. Li, Z.; Usman, S.; Zhang, J.; Zhang, Y.; Su, R.; Chen, H.; Li, Q.; Jia, M.; Amole, T.A.; Guo, X. Effects of bacteriocin-producing *Lactiplantibacillus plantarum* on bacterial community and fermentation profile of whole-plant corn silage and its in vitro rumina fermentation, microbiota, and CH_4 emissions. *J. Anim. Sci. Biotechnol.* **2024**, *15*, 107. [CrossRef] [PubMed]
38. Cai, Y.; Benno, Y.; Ogawa, M.; Ohmomo, S.; Kumai, S.; Nakase, T. Influence of *Lactobacillus* spp. from an inoculant and of *Weissella* and *Leuconostoc* spp. from forage crops on silage fermentation. *Appl. Environ. Microbiol.* **1998**, *64*, 2982–2987. [CrossRef] [PubMed]
39. Wang, Z.; Li, Q.; Li, J.; Shang, L.; Li, J.; Chou, S.; Lyu, Y.; Shan, A. pH-responsive antimicrobial peptide with selective killing activity for bacterial abscess therapy. *J. Med. Chem.* **2022**, *65*, 5355–5373. [CrossRef] [PubMed]
40. Guo, X.; Xu, D.; Li, F.; Bai, J.; Su, R. Current approaches on the roles of lactic acid bacteria in crop silage. *Microb. Biotechnol.* **2023** *16*, 67–87. [CrossRef] [PubMed]

fermentation

MDPI

Article

Effects of *Lactobacillus plantarum*-81-Fermented Feed on Growth and Intestinal Health of Muscovy Ducks

Zhaolong Li [1,*], Song Peng [1], Mengshi Zhao [1], Xiaodong Zhuang [2], Huini Wu [1], Tiecheng Sun [3] and Fengqiang Lin [1]

[1] Institute of Animal Husbandry and Veterinary Medicine, Fujian Academy of Agricultural Sciences, Fuzhou 350013, China; pengsong@faas.cn (S.P.); 13375001253@163.com (M.Z.)
[2] Fujian Zhangzhou Changlong Agriculture and Animal Husbandry Co., Ltd., Zhangzhou 363000, China
[3] Fujian Animal Husbandry Station, Fuzhou 350013, China
* Correspondence: lizhaolong@faas.cn; Tel.: +86-591-87840002

Abstract: Feed fermented by various strains of *Lactobacillus plantarum* (LP) produces distinct biologically active substances. This study systematically evaluates the growth performance, gut microbiota modulation, and immune response parameters in Muscovy ducks fed with LP81-fermented diets (LP81-FF) compared to conventional regimens. Our findings demonstrate that LP81-FF elicits dose-dependent improvements in Muscovy duck production parameters. Through a 70-day feeding trial, LP81-FF administration reduced feed intake by 3.1% and improved the average daily gain (ADG) and feed conversion ratio (FCR) by 9.18% and 6.65% ($p < 0.05$) compared to conventional feed. Systemic antioxidant capacity analysis revealed 25.99% elevation in total antioxidant capacity (T-AOC) ($p < 0.05$), accompanied by 14.37% and 30.79% increases in serum IgG and IgM levels, respectively. Immune organ indices showed dose-responsive enhancement, with the high-dose group (HD) achieving 47.27% and 28.92% increases in thymus and bursa of Fabricius indices ($p < 0.05$). Additionally, 16S rRNA sequencing revealed that LP81-FF optimized the intestinal microbial community structure of Muscovy ducks by promoting the abundance of *Bacteroides*, *Butyricicoccus*, and *Ruminococcus* (beneficial bacteria) ($p < 0.05$), while inhibiting the increase of *Escherichia-Shigella* and *Rothia* (harmful bacteria). It also promoted the secretion of beneficial metabolites such as Glutaric acid and 2,6-Diaminohexanoic acid in the intestine, inhibited the production of harmful substances dominated by Fexofenadine, and enhanced the strength of physical barrier-related factors such as intestinal mucosa villi and goblet cell count. These multi-omics insights establish that LP81-FF enhances growth performance through coordinated modulation of gut–liver axis homeostasis, mucosal immunity activation, and microbial-metabolic network optimization. Our results position LP81-FF as a sustainable alternative to antibiotic growth promoters in waterfowl production systems.

Keywords: LP81-FF; Muscovy ducks; growth performance; intestinal health

1. Introduction

Microbial-fermented feed is produced through the controlled microbial metabolism of substrates under defined temperature and moisture conditions [1]. In recent years, significant progress has been made in research on microbial applications, which has stimulated heightened scientific interest in harnessing beneficial microorganisms for the development of bioactive feed additives through fermentation processes in animal husbandry practices.

This emerging field has consequently generated a substantial body of scholarly investigation globally [2]. Probiotic-fermented feed has emerged as a promising strategy to enhance poultry health and productivity. Various probiotics, such as *Lactobacillus*, *Bacillus*, and *Saccharomyces* species, contribute distinct functional benefits during fermentation, including nutrient enrichment, pathogen inhibition, and immune modulation [3]. In poultry, these microbial additives improve gut microbiota balance, nutrient absorption, and disease resistance, thereby optimizing growth performance and feed efficiency. For chickens, studies highlight that *Lactobacillus*-fermented diets enhance intestinal integrity and weight gain, while Bacillus strains reduce oxidative stress and improve meat quality [4]. Ducks, particularly susceptible to enteric pathogens, exhibit strengthened immunity and reduced mortality when fed Saccharomyces-supplemented feed, attributed to enhanced antioxidant activity and microbial diversity [5]. Notably, Muscovy ducks, a key meat-producing species, show unique responsiveness to multi-strain probiotics, with synergistic effects on growth rates, lipid metabolism, and fecal microbiome composition. These differential responses underscore the need for species-specific probiotic formulations. Investigating tailored fermentation approaches for chickens, ducks, and Muscovy ducks could unlock precision nutrition strategies, advancing sustainable poultry production [6].

Previous studies have shown that feeding *Lactobacillus* fermented feed to livestock and poultry can cause different results for improving their production performance, specifically lowering the pH of the intestinal environment in poultry and reducing the prevalence of pathogens like *E. coli* and *Salmonella* [7]. Additionally, it also differs in enhancing nutrient digestibility, improving intestinal morphology, decreasing anti-nutritional factor levels in feed, and reducing dust and ammonia emissions in housing. Some meat ducks have been fed with LP-fermented feed to enhance the growth of small intestinal villi and increase feed intake [8].

Reports indicate that *Lactobacillus* fermentation of livestock and poultry diets produces higher levels of organic acids, releasing more small molecular substances, which contribute significantly to the maintenance of gastrointestinal balance, improving feed conversion rates (indirectly reducing production costs), and enhancing the immune response in these animals. The metabolites generated post-fermentation exhibit a range of biological activities, including antibacterial properties, antioxidant effects, mycotoxin degradation, and the decomposition of anti-nutritional factors, among other functions [9]. Furthermore, the inappropriate use of antibiotics in animal husbandry has become a growing concern in recent years. Over the past two years, raw material prices have significantly risen, leading to increased costs for waterfowl breeding and affecting the healthy development of the waterfowl breeding industry [10].

To mitigate antibiotic use, enhance feed conversion efficiency, reduce breeding costs, and ensure the stable and healthy development of the Muscovy duck industry, this study focuses on cost reduction and efficiency improvement in Muscovy duck breeding and evaluates the impact of LP81-FF on the production performance of Muscovy ducks while investigating its effects on serum biochemical immunity and other relevant factors, ultimately advancing the application of LP81-FF in Muscovy duck breeding, and conducts in-depth research on the effects of microbial-fermented feed on the production performance of Muscovy ducks. This research will provide a foundation for the application of microbial-fermented feeds in Muscovy duck breeding.

2. Materials and Methods

2.1. Animals and Dietary Treatments

A total of 1200 healthy female and male Muscovy ducks, with a weight of 43.23 ± 5.15 g, were selected and randomly divided into four groups, each containing five replicates (n = 60). Each replicate consisted of 60 ducks (we carried out pre-testing before and found that the feeding of fermented feed did not have a significant effect on the sex factor of male and female ducks. Therefore, in this project we chose the test ducks to be half male and half female) arranged in a pen. The control group (CT) was fed a basal diet (in powder form), while the experimental groups received supplements of 30% (LD), 50% (MD), and 100% (HD) LP81-FF, respectively, based on the basal diet [6]. Detailed records of their production performance and preliminary mechanisms of the effect of LP81-FF on the production performance of ducks were revealed by multi-omics and morphology methods. The test period lasted for 70 days. The LP81 strain, which was isolated in our previous study, demonstrated significant in vitro antibacterial activity (against pathogens such as *E. coli* and *Salmonella*), acid tolerance, and adhesion capacity compared to other LP strains. The basic diet is a complete feed prepared based on the nutritional requirements of Muscovy ducks from 1 to 70 days old in Nutrient Requirements of Poultry [11] and China Agricultural Industry Standard NY/T 816-2021 [12]. The specific composition and nutritional levels of the basal diet are presented in Table 1. The determination of crude protein and dry matter followed the national standards GB/T 6432-2018 [13] and GB/T 6438-2007 [14], respectively, while the determination of calcium and total phosphorus adhered to the national standards GB/T 6436-2018 [15] and GB/T 6437-2018 [16]. The metabolizable energy of the diet (MJ/kg) was calculated using the following formula: [total energy consumed (MJ)—fecal energy (MJ)]/diet intake (kg) [17]. The trial period encompassed 75 days and a 5-day pre-test phase. The animal care and use protocol was approved by the Institutional Animal Care and Use Committee at the Institute of Animal Husbandry and Veterinary Medicine of Fujian Academy Agricultural and Sciences (202402FJ026).

Table 1. Basal dietary composition and nutritional level (air-dry basis).

Items	Starter Diet (1–35 Days)	Grower Diet (35–70 Days)
Ingredients		
Corn	42.80	42.70
Soybean meal (46% CP)	23.30	22.10
Wheat bran	21.1	22.12
Soybean oil	2.00	0
Limestone	1.15	1.10
$CaHPO_4$	1.17	1.17
NaCL	0.48	0.81
Rice	6.00	8.00
Premix [1]	2.00	2.00
Total	100.00	100.00
Nutrient levels [2]		
ME/(MJ/kg)	2.62	2.12
DM	83.57	84.56
CP	18.37	16.85
Ca	1.21	1.25
AP	0.29	0.30
Met	0.51	0.48
Lys	1.01	0.91
Try	0.19	0.20
Thr	0.68	0.49

(1) The premix provided the following per kg of diets: VA 11900IU, VD3 3630 IU, VE 20IU, VK3 2.7 mg, VB1 2.0 mg, VB2 8.7 mg, D-pantothenic acid 12.3 mg, Nicotinic acid 59.0 mg, VB6 4.8 mg, VB12 35 μg, Biotin 0.18 mg, Folic acid 2.1 mg, Fe (as ferrous sulfate) 35.0 mg, Cu (as copper sulfate) 3.6 mg, Zn (as zinc sulfate) 35.7 mg, Mn (as manganese sulfate) 46.0 mg, I (as potassium iodide) 0.28 mg, Se (as sodium selenite) 0.20 mg. (2) ME was a calculated value, while the others were measured values.

2.2. Preparation of LP-81-Fermented Feed

The LP81 strain, isolated from Tibetan kimchi in 2021 and maintained by this research center, contains 1.45×10^8 CFU/mL of viable bacteria. The ratio of LP81 solution to water and complete feed for fermented feed preparation is in the ratio of 3:180:1000. The mixture was stirred thoroughly and placed into a fermentation tank, which was then sealed and incubated at 35 °C for 24 h to produce fermented feed. Fresh batches of fermented feed were prepared daily and were readily available for use.

2.3. Feeding and Management Test

Muscovy ducks were fed using an ad libitum feeding regimen. A total of 1200 experimental ducks were raised in a single building, using a mesh bed breeding system. The ambient temperature for young Muscovy ducks was maintained at 35 °C, while the temperature for mature Muscovy ducks was set at 25 °C. Immunization, deworming, health care, and disinfection were conducted in accordance with the routine procedures established by the duck farm.

2.4. Sample Collection

2.4.1. Blood

A total of 5 mL of blood was collected from the vein. After allowing the samples to stand at room temperature for 30 min, they were centrifuged at 3000× *g* r/min for 20 min. The supernatant was then carefully removed and stored at −20 °C for subsequent analysis.

2.4.2. Intestinal Content and the Muscle Tissue

Ten Muscovy ducks from each group were randomly selected and euthanized under anesthesia. The entire duodenum, jejunum, ileum, cecum, and a 2 cm segment of the cecum were collected, and the thymus, bursa of fabricius, and spleen were subsequently collected. The intestinal tissues were fixed in 4% paraformaldehyde, followed by dehydration, clearing, and embedding in wax. Subsequently, the samples were sliced using a microtome for future use and stored at −4 °C.

2.5. Indicator Determination

2.5.1. Determination of the Nutrient Content of the Feed Before and After Fermentation

The dry matter content of the feed, both before and after fermentation, determined in accordance with GB/T 6435-2006 [18]. The crude protein content was assessed based on GB/T 6432-2018, while the crude fiber content was evaluated following GB/T 6434-2022 [19]. The crude ash content was referenced from GB/T 6438-2007, and the total phosphorus and calcium contents were respectively determined according to GB/T 6437-2018 and GB/T 6436-2018. For the analysis, 25 g of the feed was taken both before and after fermentation, to which 225 mL of sterile saline was added. The mixture was then thoroughly mixed, allowed to stand, and the pH value was measured using a pH meter.

2.5.2. Measurement of the Growth Performance

On the mornings of the 1st, 35th, and 70th days of the experiment, Muscovy ducks were weighed, and their body weights were recorded. Throughout the study, the average daily feed intake per group and per animal (ADFI) was documented and calculated. At the end of experiment, the weights of both the experimental Muscovy ducks and the control group were measured. Subsequently, the average daily gain (ADG) and feed-to-weight ratio (F/G) for the four groups of Muscovy ducks were calculated. The formulas used for these calculations are as follows: ADG (kg/d) = (last weight − initial weight)/number of test days; ADFI (kg/d) = total feed intake/number of test days; F/G = ADFI/ADG.

2.5.3. Determination of Blood Biochemistry, Immune, and Antioxidant Indicators in Muscovy Ducks

A fully automated blood biochemistry analyzer (Beckman AU51200, Miami, FL, USA) was utilized to measure total protein (TP), albumin (TPT), glucose (GLU), urea nitrogen (UN), urea (U), uric acid (UA), triglycerides (TGs), total cholesterol (TC), high-density lipoprotein (HDL), and low-density lipoprotein (LDL). Blood immune indicators, including immunoglobulin G (IgG), immunoglobulin M (IgM), and immunoglobulin A (IgA), were obtained from Tiangen Biotechnology (Beijing, China) Co., Ltd. (Beijing, China). Additionally, antioxidant capacity indicators, such as malondialdehyde (MDA), total antioxidant capacity (T-AOC), total superoxide dismutase (T-SOD), catalase (CAT), and glutathione peroxidase (GSH-Px), were sourced from Shanghai Sangon Bioengineering Technology Services Co., Ltd. (Shanghai, China). The determination methods were performed in accordance with the kit operating instructions.

2.5.4. Immune Organ Index Measurement

The thymus, bursa of fabricius, and spleen were subsequently collected. Excess water and blood were absorbed using filter paper, after which the samples were weighed and analyzed. The organ index was calculated using the formula Immune Organs Index (mg/g) = immune organ weight (mg)/live body weight (g).

The paraffin sections of the previously prepared intestinal tissues were deparaffinized in a sequential manner. Following conventional Hematoxylin-Eosin (HE) staining, dehydration and gum sealing were performed. The general structure of the intestines in each group was observed under a 100× light microscope, while the morphology and distribution of the intestinal mucosa, submucosa, muscle layer, and serosa were examined under a 400× light microscope. Subsequently, DigiLab-C software 3.0 was utilized to measure villus height and crypt depth. For each slice, five fields of view were selected, and the average was calculated as the final result; data were recorded, and villus height/crypt depth was computed. Additionally, paraffin sections of the intestinal tissues from each group were randomly selected from five fields of view at 400× magnification, and goblet cells were counted using DigiLab-C software. The goblet cells in each group were counted three times, and statistical analysis was conducted using SPSS 19.0. The software processed the data, calculated the averages, and assessed whether the differences between each group were significant, as well as analyzed changes in the number of goblet cells.

2.6. 16S rDNA Amplicon Sequencing Method for Duodenum Contents

The primary sequencing process involves several crucial steps: extraction of total microbial DNA from intestinal contents, PCR amplification, product purification, library preparation and quality control, and sequencing on a sequencing machine. Firstly, the total microbial DNA is extracted from the intestinal contents using appropriate extraction kits and protocols. This step ensures that the DNA of interest, specifically the 16S rDNA, is isolated for further analysis. Secondly, PCR amplification is performed to generate sufficient amounts of the target 16S rDNA region. Specific primers are used to amplify the variable regions of the 16S rDNA gene, which contain the phylogenetic information needed for species identification and diversity analysis. After PCR amplification, the products are purified to remove any contaminants or unincorporated primers. This purification step is crucial for ensuring the quality of the sequencing library. Next, the sequencing library is prepared by adding adapters and barcodes to the purified PCR products. The library is then subjected to quality control to ensure its suitability for sequencing. Once the library is ready, it is loaded onto a sequencing machine for sequencing. This process generates raw sequencing data that contain information about the microbial community present in the intestinal contents. Following sequencing, the raw data are processed using overlap-based assembly to stitch together the sequencing reads. Quality control measures, such as chimera filtering, are applied to remove any low-quality or artifactual sequences, resulting in high-quality clean data. Subsequently, the concept of Amplicon Sequence Variants (ASVs) is employed to construct an ASV table, which is analogous to an OTU (Operational Taxonomic Unit) table. The ASV table and representative ASV sequences are obtained, allowing for further analysis, such as diversity analysis, taxonomic annotation, and differential analysis.

2.7. *Metabolite Measurement Method for Duodenum Contents*

Duodenal content samples are analyzed using a Vanquish ultra-performance liquid chromatography (UPLC) system from Thermo Fisher Scientific (Waltham, MA, USA). Chromatographic separation of sample compounds is achieved using a Waters ACQUITY UPLC BEH Amide column (2.1 mm × 50 mm, 1.7 μm). The mobile phases employed in the liquid chromatography consist of an aqueous phase (A), containing 25 mmol/L ammonium acetate and 25 mmol/L ammonia solution, and acetonitrile (B). The sample tray temperature is maintained at 4 °C, with an injection volume of 2 μL. For mass spectrometry analysis, an Orbitrap Exploris 120 mass spectrometer (Thermo Fisher, Waltham, MA, USA) controlled by Xcalibur software (version 4.4, Thermo) is utilized. This mass spectrometer has the capability to acquire both first-order and second-order mass spectral data. Raw data obtained from the mass spectrometer are converted into mzXML format using ProteoWizard software (3.0.9134). Subsequently, metabolite identification is performed using the BiotreeDB (V3.0) database. Following metabolite identification, visual analysis is conducted to interpret the data. To identify differential metabolites, both univariate and multivariate statistical analyses are performed. Specifically, the Student's *t*-test is employed to screen for differential metabolites with a significance level of $p < 0.05$. Additionally, variables with a Variable Importance in the Projection (VIP) value greater than 1 from the first principal component of the Orthogonal Partial Least Squares Discriminant Analysis (OPLS-DA) model are further considered as differential metabolites.

2.8. *Pearson Correlation Analysis Using Spearman's Correlation Coefficient*

To evaluate correlations between 30 duodenal mucosa microbial species and various physiological, metabolic parameters (mucosal-associated, blood antioxidant, immune organ indices, seven key intestinal metabolites, and production performance), we utilized Pearson correlation analysis with Spearman's correlation coefficient. In this analysis, microbial species served as independent variables, while the other parameters were dependent. We calculated Spearman's ρ for each variable pair, ranging from −1 (perfect negative correlation) to 1 (perfect positive correlation), with 0 indicating no correlation. Statistical tests determined significance (p-value < 0.05 considered significant). This approach aimed to identify microbial species significantly correlated with the mentioned parameters, offering insights into their potential roles in colonic health, antioxidant status, immune function, metabolism, and production performance.

2.9. *Statistical Analysis*

The test data were analyzed using one-way analysis of variance (ANOVA) with SPSS version 19.0, and a *t*-test was employed for group comparisons. The results are presented as mean ± standard error (SE). A p-value greater than 0.05 indicates that the difference is not significant, while a p-value less than 0.05 indicates that the difference is significant.

3. Results

3.1. Effect of LP81-FF on Nutritional Components of Feed

Table 2 illustrates that following fermentation of the complete feed by LP-81, there was a noticeable decline in dry matter content. In contrast, the levels of crude protein, crude fiber, crude ash, calcium, and phosphorus exhibited an upward trend, although the differences were not statistically significant. Additionally, the pH value experienced a significant reduction ($p < 0.05$).

Table 2. Nutrient variation of complete feed before and after fermentation (air-dry basis).

Items	Before Fermentation		After Fermentation	
Time (D)	1–35 days	35–70 days	1–35 days	35–70 days
DM/%	83.57 ± 0.23	84.56 ± 0.41	82.53 ± 0.51	83.21 ± 0.29
TP/%	18.37 ± 0.15	16.85 ± 0.03	18.56 ± 0.26	16.11 ± 0.17
CF/%	5.01 ± 0.08	6.05 ± 0.31	4.97 ± 0.03	5.89 ± 0.06
Ash/%	8.29 ± 0.11	9.30 ± 0.32	8.33 ± 0.05	9.17 ± 0.01
Ca (%)	1.21 ± 0.14	1.25 ± 0.21	1.25 ± 0.09	1.34 ± 0.15
TP %	0.47 ± 0.03	0.55 ± 0.01	0.50 ± 0.02	0.58 ± 0.02
pH	6.65 ± 0.18 ^{a}	6.67 ± 0.03 ^{a}	5.13 ± 0.16 ^{b}	5.27 ± 0.17 ^{b}

Note: Values are mean ± SE, $n = 10$. The different lowercase letters in the table are significant ($p < 0.05$), and the same or no letters are not significant ($p > 0.05$). The following tables are the same.

3.2. Effect of LP81-FF on the Growth Performance of Muscovy Ducks

As shown in Table 3, there were no significant differences in initial weight, final weight, or daily weight gain between the feed fermentation group and the control group on the 35th day of the experiment ($p > 0.05$). However, the average daily feed intake was significantly reduced ($p < 0.05$). Notably, the average daily feed intake of the HD was 3.59 g lower than that of CT, and this difference was significant ($p < 0.05$). Additionally, the material-to-weight ratio exhibited a significant decrease ($p < 0.05$). On the 70th day of the experiment, through a 70-day feeding trial, LP81-FF administration reduced feed intake by 3.1% and improved the average daily gain (ADG) and feed conversion ratio (FCR) by 9.18% and 6.65% ($p < 0.05$) compared to conventional feed, with MD and HD showing the most pronounced reductions ($p < 0.05$). As a result, the cost of farming is significantly reduced.

Table 3. Effects of LP-81-fermented diet on the production performance of Muscovy ducks.

Days	Items	CT	LD	MD	HD	p-Value
35	Initial weight/g	43.56 ± 3.12	43.11 ± 5.17	43.25 ± 4.17	43.58 ± 4.29	0.563
	Final weight/g	797.17 ± 5.69	799.12 ± 9.36	792.05 ± 10.36	789.05 ± 8.57	0.097
	ADG/g	21.53 ⊥ 3.36	21.60 ± 4.68	21.39 ± 3.29	21.30 ± 6.21	0.055
	ADFI/g	38.67 ± 5.54 ^{a}	37.95 ± 5.09 ^{b}	36.24 ± 7.21 ^{b}	35.08 ± 4.31 ^{b}	0.013
	F/G	1.80 ± 0.13 ^{a}	1.74 ± 0.17 ^{b}	1.69 ± 0.13 ^{c}	1.65 ± 0.15 ^{c}	0.024
70	Initial weight/g	797.17 ± 5.69	799.12 ± 9.36	792.05 ± 10.36	789.05 ± 8.57	0.097
	Final weight/g	1731.22 ± 5.71	1792.33 ± 6.05	1805.54 ± 6.37	1809.01 ± 7.06	0.089
	ADG/g	26.69 ± 8.14	28.38 ± 6.05	28.96 ± 5.12	29.14 ± 5.50	0.054
	ADFI/g	100.35 ± 11.16 ^{b}	103.30 ± 10.75 ^{a}	103.10 ± 9.93 ^{a}	102.28 ± 10.1 ^{a}	0.012
	F/G	3.76 ± 0.15 ^{a}	3.64 ± 0.12 ^{b}	3.56 ± 0.10 ^{c}	3.51 ± 0.13 ^{c}	0.011

Note: Values are mean ± SE, $n = 10$. The different lowercase letters in the table are significant ($p < 0.05$), and the same or no letters are not significant ($p > 0.05$). The following tables are the same.

3.3. Effect of LP81-FF on Biochemical Parameters of Muscovy Ducks

Table 4 demonstrates that fermented feed can significantly enhance the total protein content ($p < 0.05$) and increase the levels of albumin and alkaline phosphatase in serum

($p < 0.05$). Notably, the serum albumin and alkaline phosphatase levels in the MD and HD were significantly higher than those observed in the CT and the LD ($p < 0.05$). However, no significant differences were found in other serum biochemical indicators ($p > 0.05$).

Table 4. Improvement of serum biochemical indexes of Muscovy ducks by LP-81-fermented feeds.

Items	The Proportion of LP-Fermented Feed %				*p*-Value
	CT	LD	MD	HD	
GLU/(mmol/L)	5.65 ± 0.23	6.03 ± 0.55	6.08 ± 0.04	6.35 ± 0.24	0.179
TP/(g/L)	33.25 ± 4.13 ^{b}	42.55 ± 8.90 ^{a}	45.12 ± 7.53 ^{a}	46.30 ± 7.16 ^{a}	0.033
ALB/(g/L)	10.15 ± 1.27 ^{c}	12.15 ± 1.53 ^{b}	13.94 ± 3.21 ^{a}	14.20 ± 3.65 ^{a}	0.041
GLO/(g/L)	23.15 ± 1.25	24.36 ± 3.59	24.27 ± 6.00	22.15 ± 3.45	0.167
A/G	0.55 ± 0.13	0.61 ± 0.34	0.57 ± 0.12	0.58 ± 0.19	0.105
ALP/(U/L)	175.13 ± 21.33 ^{c}	189.60 ± 27.65 ^{b}	207.56 ± 28.43 ^{a}	208.20 ± 35.14 ^{a}	0.035
UN/(mmol/L)	0.49 ± 0.02	0.73 ± 0.04	0.63 ± 0.02	0.65 ± 0.02	0.371
UA/(mmol/L	103.34 ± 14.57	110.41 ± 8.21	105.23 ± 11.37	106.33 ± 10.04	0.071
TG/(mmol/L)	0.66 ± 0.04	0.69 ± 0.01	0.65 ± 0.03	0.61 ± 0.04	0.017
TC/(mmol/L)	3.33 ± 0.42	3.27 ± 0.15	3.37 ± 0.23	3.25 ± 0.14	0.121
HDL/(mmol/L)	1.35 ± 0.13	1.38 ± 0.21	1.32 ± 0.50	1.37 ± 0.22	0.098
LDL/(mmol/L)	1.30 ± 0.17	1.31 ± 0.23	1.35 ± 0.23	1.33 ± 0.53	0.155

Note: Values are mean ± SE, $n = 10$. The different lowercase letters in the table are significant ($p < 0.05$), and the same or no letters are not significant ($p > 0.05$). The following tables are the same.

3.4. *Effect of LP81-FF on Immune Index and Antioxidant Capacity of Muscovy Ducks*

The liver index showed no significant differences among the CT, LD, MD, and HD groups (0.47–0.55; $p = 0.582$). In contrast, the thymus index increased significantly in the HD group (4.05 ± 0.31) compared to CT (2.75 ± 0.32), LD (3.57 ± 0.06), and MD (3.68 ± 0.32) ($p = 0.038$), with distinct alphabetical superscripts indicating intergroup differences. Similarly, the Fabricius index was significantly higher in the HD group (1.56 ± 0.23) than in CT (1.21 ± 0.01), LD (1.41 ± 0.21), and MD (1.39 ± 0.15) ($p = 0.024$), demonstrating dose-dependent enhancement. These results suggest that LP-81-fermented feed positively influenced immune organ indices (thymus and Fabricius) in a dose-responsive manner, while no effect was observed on the liver index (Table 5).

Table 5. Enhancement of immune organ indexes in Muscovy ducks by LP-81-fermented feed.

Items	CT	LD	MD	HD	*p*-Value
Liver index	0.47 ± 0.01	0.49 ± 0.02	0.53 ± 0.06	0.55 ± 0.05	0.582
Thymus index	2.75 ± 0.32 ^{c}	3.57 ± 0.06 ^{b}	3.68 ± 0.32 ^{b}	4.05 ± 0.31 ^{a}	0.038
Fabricius index	1.21 ± 0.01 ^{c}	1.41 ± 0.21 ^{b}	1.39 ± 0.15 ^{b}	1.56 ± 0.23 ^{a}	0.024

Note: Values are mean ± SE, $n = 10$. The different lowercase letters in the table are significant ($p < 0.05$), and the same or no letters are not significant ($p > 0.05$). The following tables are the same.

Table 6 demonstrates that the use of fermented feed significantly showed dose-responsive, enhancement-immune organ indices, with the high-dose group (HD) achieving 47.27% and 28.92% increases in the thymus and bursa of Fabricius indices ($p < 0.05$). Furthermore, the levels of immunoglobulin G (IgG) and immunoglobulin A (IgM) in the serum of the fermented feed group increased 14.37% and 30.79%, which were significantly higher than those in the control group ($p < 0.05$), with MD and HD exhibiting the most substantial increases. Notably, there was no significant difference in the levels of immunoglobulin A (IgA) among the four groups ($p > 0.05$). Additionally, Muscovy ducks fed with LP81-FF demonstrated 25.99% elevation in total antioxidant capacity (T-AOC) ($p < 0.05$), malondialdehyde (MDA), and superoxide dismutase (SOD), with significant differences observed ($p < 0.05$). The highest levels of MDA and SOD were recorded in HD, showing significant differences ($p < 0.05$). However, no significant differences were found among the four groups regarding catalase (CAT) and glutathione peroxidase (GSH-Px) levels in the Muscovy ducks' serum ($p > 0.05$).

Table 6. Enhancement of immune and antioxidant indices in Muscovy ducks by LP-81-fermented feed.

Items	CT	LD	MD	HD	*p*-Value
IgG/(mg/mL)	43.15 ± 5.36 ^{c}	45.21 ± 3.46 ^{b}	47.08 ± 3.65 ^{b}	49.35 ± 4.06 ^{a}	0.047
IgM/(mg/mL)	3.15 ± 0.21 ^{c}	3.47 ± 0.29 ^{b}	3.99 ± 0.25 ^{b}	4.12 ± 0.15 ^{a}	0.031
IgA/(mg/mL)	0.38 ± 0.05	0.35 ± 0.01	0.39 ± 0.04	0.43 ± 0.05	0.562
T-AOC/(U/mL	3.27 ± 0.36 ^{c}	3.99 ± 0.21 ^{b}	4.01 ± 0.31 ^{a}	4.12 ± 0.27 ^{a}	0.013
MDA(nmol/mL)	3.05 ± 0.29 ^{c}	3.89 ± 0.17 ^{b}	3.99 ± 0.33 ^{b}	4.25 ± 0.37 ^{a}	0.019
CAT(U/mL)	2.75 ± 0.03	2.85 ± 0.31	2.77 ± 0.61	2.85 ± 0.12	0.314
SOD/(U/mL)	40.12 ± 5.36 ^{c}	53.22 ± 4.98 ^{b}	54.55 ± 8.16 ^{b}	58.34 ± 9.67 ^{a}	0.028
GSH-Px/(U/mL	344.21 ± 18.53	337.35 ± 28.26	355.11 ± 18.52	357.15 ± 28.65	0.376

Note: Values are mean ± SE, $n = 10$. The different lowercase letters in the table are significant ($p < 0.05$), and the same or no letters are not significant ($p > 0.05$). The following tables are the same.

3.5. Effect of LP81-FF on Intestinal Morphological Structure of Muscovy Ducks

Figure 1 and Table 7 illustrates that, compared to the control group, the villous height in the jejunum and cecum of the group fed LP81-FF was significantly higher ($p < 0.05$). In contrast, the crypt height of the ileum in Muscovy ducks receiving the fermented feed showed a decrease when compared to the control group, with a significant difference noted ($p < 0.05$). Furthermore, the ratios of villus height to crypt depth in the jejunum and cecum of the LP81-FF group exhibited a significant increase ($p < 0.05$). However, the villus-to-crypt (V/C) ratios in the duodenum and ileum across all groups did not show significant differences ($p > 0.05$).

Table 7. Improvement of intestinal morphological structure of Muscovy ducks by LP81-FF.

Items	CT	LD	MD	HD	*p*-Value
Duodenum					
Villus height/μm	312.23 ± 5.75	309.13 ± 8.45	316.30 ± 9.61	305.42 ± 7.36	0.156
Crypt depth/μm	157.21 ± 5.76	155.36 ± 7.85	143.37 ± 8.05	151.33 ± 7.06	0.235
V/C	1.45 ± 0.12	1.47 ± 0.25	1.51 ± 0.24	1.45 ± 0.15	0.452
Jejunum					
Villus height/μm	345.61 ± 10.76 ^{c}	355.30 ± 8.23 ^{b}	375.51 ± 5.11 ^{a}	378.44 ± 6.73 ^{a}	0.013
Crypt depth/μm	180.17 ± 5.33	182.45 ± 6.18	183.56 ± 7.08	183.45 ± 6.55	0.057
V/C	1.51 + 0.14 ^{b}	1.57 + 0.21 ^{b}	2.07 + 0.77 ^{a}	2.06 + 0.32 ^{a}	0.043
Ileum					
Villus height/μm	255.31 ± 5.60	257.35 ± 6.31	259.32 ± 1.27	252.13 ± 5.02	0.152
Crypt depth/μm	131.21 ± 3.55 ^{b}	133.54 ± 7.06 ^{a}	137.21 ± 2.31 ^{a}	135.11 ± 2.15 ^{a}	0.023
V/C	1.55 ± 0.06 ^{b}	1.53 ± 0.12 ^{b}	1.89 ± 0.21 ^{a}	1.86 ± 0.47 ^{a}	0.013
Cecum					
Villus height/μm	103.22 ± 2.39 ^{c}	135.24 ± 3.25 ^{b}	138.54 ± 6.17 ^{a}	139.24 ± 5.56 ^{a}	0.015
Crypt depth/μm	65.23 ± 5.89	80.12 ± 6.31	81.22 ± 3.22	82.55 ± 4.20	0.057
V/C	1.58 ± 0.11 ^{b}	1.69 ± 0.21 ^{a}	1.70 ± 0.32 ^{a}	1.69 ± 0.37 ^{a}	0.014

Note: Values are mean ± SE, $n = 10$. The different lowercase letters in the table are significant ($p < 0.05$), and the same or no letters are not significant ($p > 0.05$). The following tables are the same.

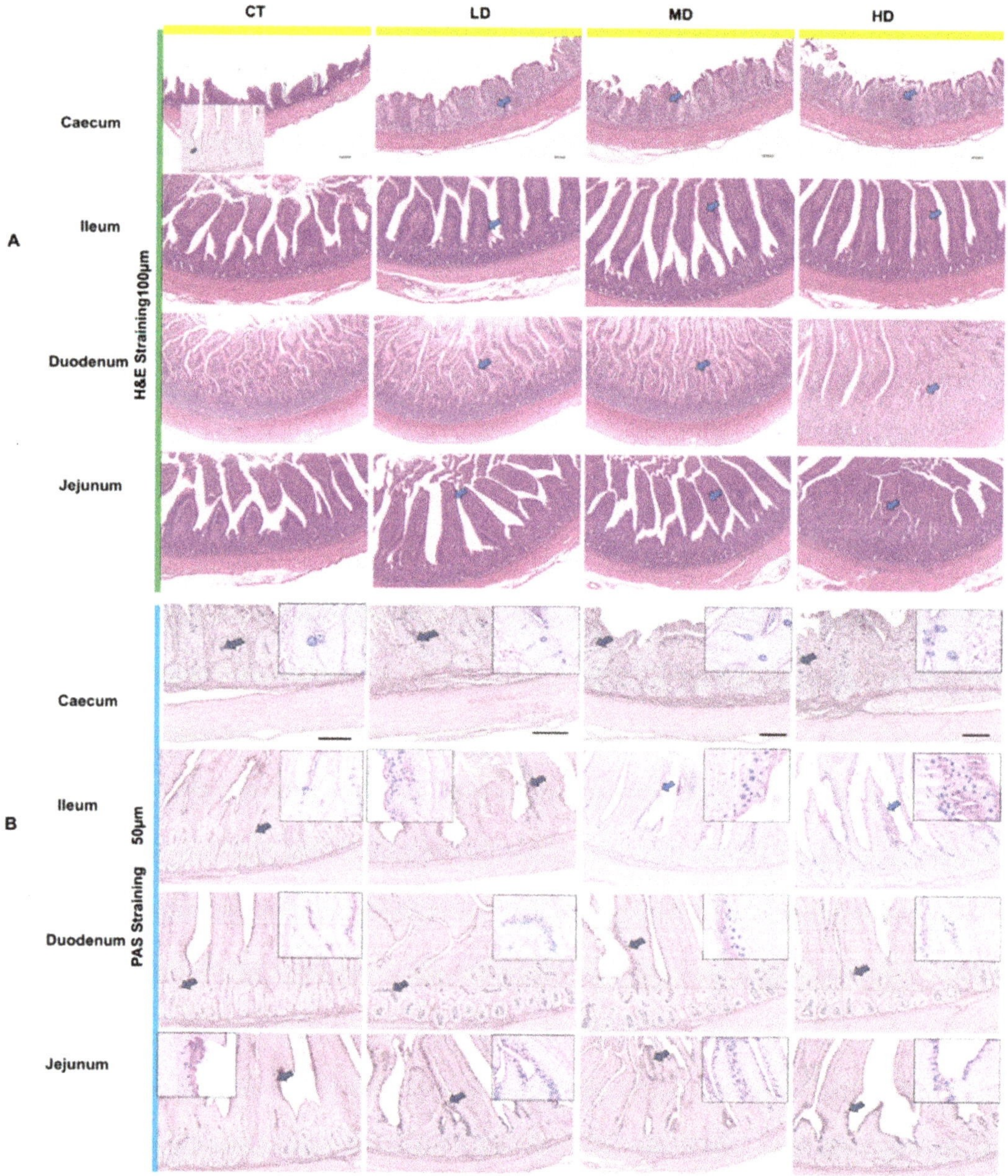

Figure 1. Histopathological analysis of the intestine using H&E staining and PAS staining. (**A**) H&E staining of the ileum, duodenum, jejunum, and cecum (100 μm). (**B**) PAS staining of the ileum, duodenum, jejunum, and cecum (50 μm). A blue arrow indicates the significant differences between the LP81-FF group and CT groups.

3.6. *Effect of LP81-FF on the Intestinal Mucosal Status of Muscovy Ducks*

Goblet cells are a typical type of mucus-secreting cell, primarily composed of mucin. Mucin particles are secreted extracellularly, and synthesized mucus, including water and inorganic salts, adheres to the surface of the intestinal mucosa, forming a protective and lubricating intestinal mucus layer(Figure 1). This layer is involved in the occurrence and development of various intestinal diseases. As shown in Table 8, feeding fermented feed significantly increased the number of goblet cells on the intestinal mucosal surface of the ileum and cecum in Muscovy ducks compared to the control group. Notably, the intestinal mucosal surfaces of the jejunum and cecum in the LP81-FF group exhibited a significant increase in goblet cell numbers. The difference in the number of goblet cells between this group and the control group was extremely significant ($p < 0.01$). Additionally, while there was an increasing trend in the number of goblet cells on the intestinal mucosal surface of the ileum in the LP81-FF group compared to the control, the increase was more pronounced in the 100% supplemented group, although this difference was not statistically significant ($p > 0.05$). Furthermore, there was no significant difference in the number of goblet cells on the duodenal intestinal mucosal surface among the groups ($p > 0.05$).

Table 8. Effect of LP-fermented feed on goblet cells in intestinal mucosa of Muscovy ducks.

Items	CT	LD	MD	HD	*p* Value
Duodenum	101.12 ± 2.65^{c}	116.11 ± 3.57^{c}	131.11 ± 5.23^{b}	143.56 ± 8.21^{a}	0.018
Jejunum	93.31 ± 4.99^{c}	106.72 ± 8.23^{b}	126.37 ± 3.59^{a}	131.54 ± 9.12^{a}	0.011
Ileum	71.57 ± 5.63^{d}	105.21 ± 5.69^{c}	135.56 ± 6.13^{b}	167.83 ± 3.32^{a}	0.017
Cecum	13.70 ± 2.57^{b}	18.57 ± 3.76^{b}	18.01 ± 1.22^{b}	25.34 ± 1.75^{a}	0.048

Note: Values are mean ± SE, $n = 10$. The different lowercase letters in the table are significant ($p < 0.05$), and the same or no letters are not significant ($p > 0.05$). The following tables are the same.

3.7. Impact of LP81-FF on the Duodenum Microbiota of Muscovy Ducks

The influence of LP81-FF on the genus-level microbiota in the duodenum of Muscovy ducks is depicted in Figure 2A,B. At the genus level, the most abundant taxa in the ducks' duodenum were *Faecalibacterium, Ligilactobacillus, Enterococcus, Bacteroides, Streptococcus, Rothia, Lachnospiraceae_ unclassified, Chloroplast_unclassified, Clostridia_UCG-014_unclassified, Escherichia-Shigella, Ruminococcaceae_ unclassified.* Among these, significant differences were observed in *Streptococcus, Enterococcus, Chloroplast_unclassified, Escherichia-Shigella, Ligilactobacillus, Bacteroides, Erysipelatoclostridium,* and *Candidatus_Arthromitus.* The LP-81-fermented feed significantly increased the abundance of *Streptococcus, Enterococcus, Chloroplast_unclassified, Ligilactobacillus,* and *Bacteroides.* Notably, the upregulation of *Bacteroides* was most pronounced in the HD group. Conversely, in the CT group, harmful bacteria such as *Rothia, Clostridia_UCG-014_unclassified,* and *Escherichia-Shigella* were also significantly upregulated, with *Escherichia-Shigella* showing the most significant increase (Figure 2C).

LefSe analysis was conducted to highlight the differences in microbiota between different treatment groups. As shown in Figure 2E,F, significant alterations in the microbial community were observed. In the HD group, the differential taxa were mainly concentrated in 13 bacterial groups ($p < 0.05$), including Ruminococcaceae_unclassified, Bacteroides, and Enterococcus. In the MD group, the differential taxa were primarily focused in five bacterial groups, namely *Ligilactobacillus, Enterococcus,* and *f-lactobacilaceae.* In the LD group, the differential taxa were mainly concentrated in six bacterial groups, including *g__Streptococcus, f__Streptococcaceae, s__Streptococcus__equines,* and *f__Lactobacillaceae.* The differential microbiota in the CT group were primarily centered on *g__Escherichia-Shigella* and *g__Candidatus_Arthromitus* ($p < 0.05$). These results indicate that compared to the control group, feeding with LP81-FF significantly altered the colonic microbiota structure of Muscovy ducks.

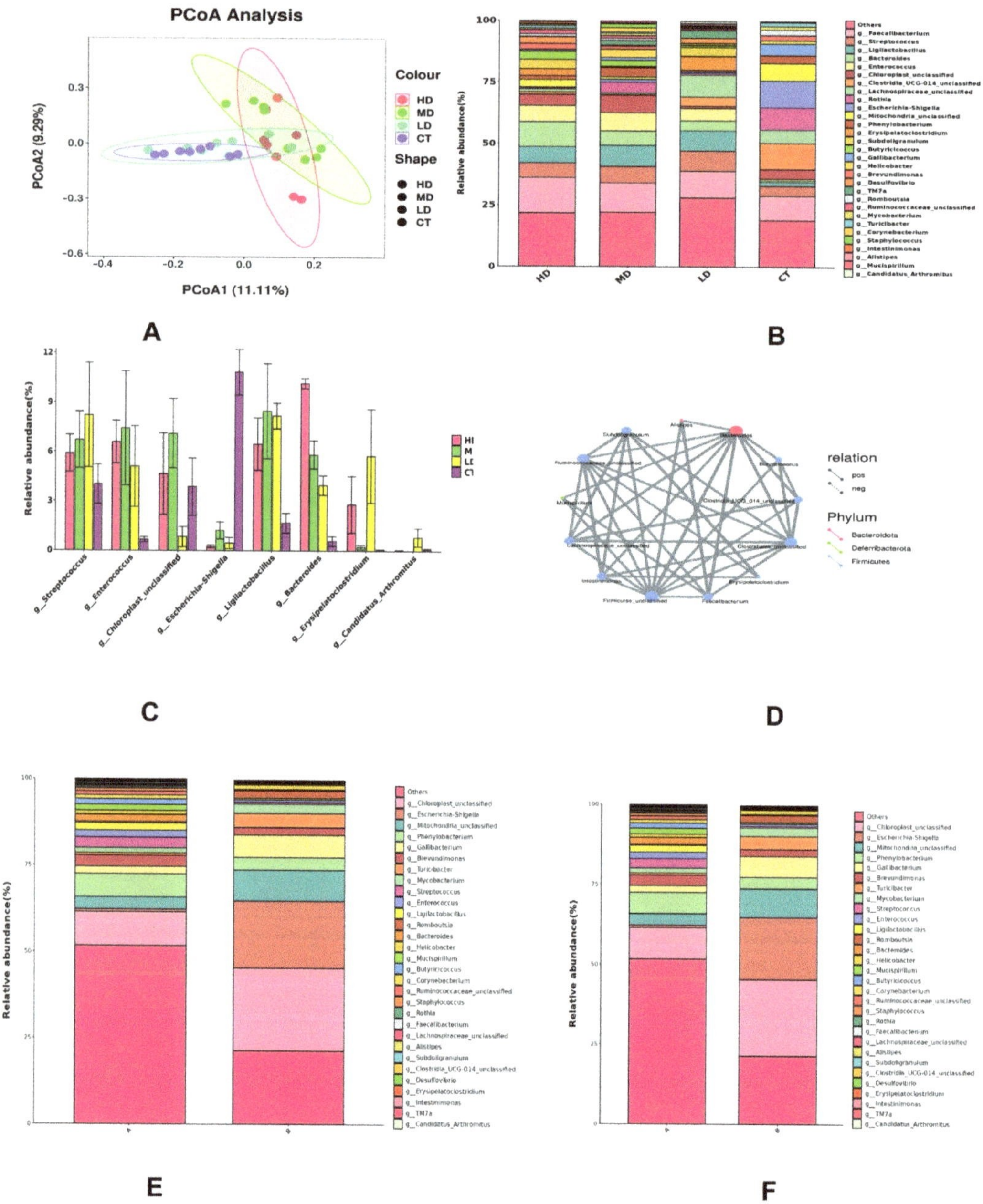

Figure 2. Effects of LP81-fermented feed on the microorganisms of the duodenum in Muscovy ducks. (**A**) Beta diversity of duodenal microbiota, with different colors representing various groupings. (**B**) Genus-level microbes exhibiting distinct differences in abundance. (**C**) Based on the microbial species abundance table and the annotation table (genus), the abundance of the top 30 species classifications was selected. (**D**) Intestinal microbiota with a high association rate. (**E**,**F**) LEfSe (LDA Effect Size) analysis identified species with significant differences in abundance between the various groups.

3.8. Identification of Core Gut Microbiota in Muscovy Ducks

The dominant species identified among the four groups at the phylum level encompassed 13 core gut microbiota, *including Ruminococcaceae_ unclassified, Firmicutes, Bacteroidota, Firmicutes_unclass, Clostridiales_unclassified, Erysipelatoclostridium, Lachnospiraceae_ unclassified, Faecalibacterium*, and *Butyricicoccus*. Among these, the most core genera included *Firmicutes, Bacteroidota, Firmicutes_unclass, Clostridiales_unclassified, and Ruminococcaceae_unclassified* (Figure 2D). These findings highlight the significant composition of the gut microbiota in Muscovy ducks, with particular emphasis on these core taxa.

3.9. Effect of Duck-Derived Composite Bacterial Additive on Intestinal Metabolites

Based on the LC-MS platform and a self-constructed database, a total of 32,722 compounds were detected in this metabolomics sequencing analysis. Among them, the most significant differences between the HD group and the CT group were observed for Formiminoglutamic acid, Kaempferol, 2,6-diaminohexanoic acid, Methylimidazoleacetic acid, Glyceraldehyde, and Xanthosine ($p < 0.05$). Additionally, significant differences were also noted for Tryptophanamide and Gamma-glutamyl-L-putrescine. These findings suggest that the duck-derived composite bacterial additive has a profound impact on the intestinal metabolites of the ducks (Figure 3A–E).

3.10. Correlation Analysis

Based on Pearson correlation analysis, the study compared the correlations between 30 microbial species in the intestinal mucosa with significant abundance changes after treatment with LP81-FF and various parameters, including intestinal mucosa-associated indices, blood antioxidant indicators, immune organ indices, seven major differential metabolites in intestinal contents, and production performance. The results revealed that the most significantly increased microbial species after feeding with LP-81-fermented feed, namely Bacteroides, Butyricicoccus, and Ruminococcus, showed positive correlations with the villus length and goblet cell count in the jejunum and ileum, as well as with the concentrations of ALB, ALP, IgM, and IgG in peripheral blood. These microbial species were also positively correlated with intestinal metabolites such as Glutaric acid and 2,6-Diaminohexanoic acid and negatively correlated with the feed-to-meat ratio at 30 and 70 days. Conversely, the numbers of harmful bacteria such as *Escherichia-Shigella, Gallibacterium, Rothia*, and *Clostridia UCG-014* in Muscovy ducks fed with LP81-FF decreased significantly and were negatively correlated with intestinal mucosa-associated indices, blood antioxidant indicators, immune organ indices, and intestinal contents ($p < 0.05$). These findings confirm that LP81-FF promotes the abundance of beneficial intestinal bacteria such as Bacteroides, Butyricicoccus, and Ruminococcus, inhibits the abundance of harmful bacteria such as Escherichia-Shigella and Rothia, increases beneficial metabolites such as short-chain fatty acids like Glutaric acid and 2,6-Diaminohexanoic acid in the intestine, reduces harmful substances dominated by Fexofenadine, enhances the strength of physical barrier-related factors such as intestinal mucosa villi and goblet cell count, lowers the feed-to-meat ratio, and ultimately promotes enhanced production performance (Figure 3F).

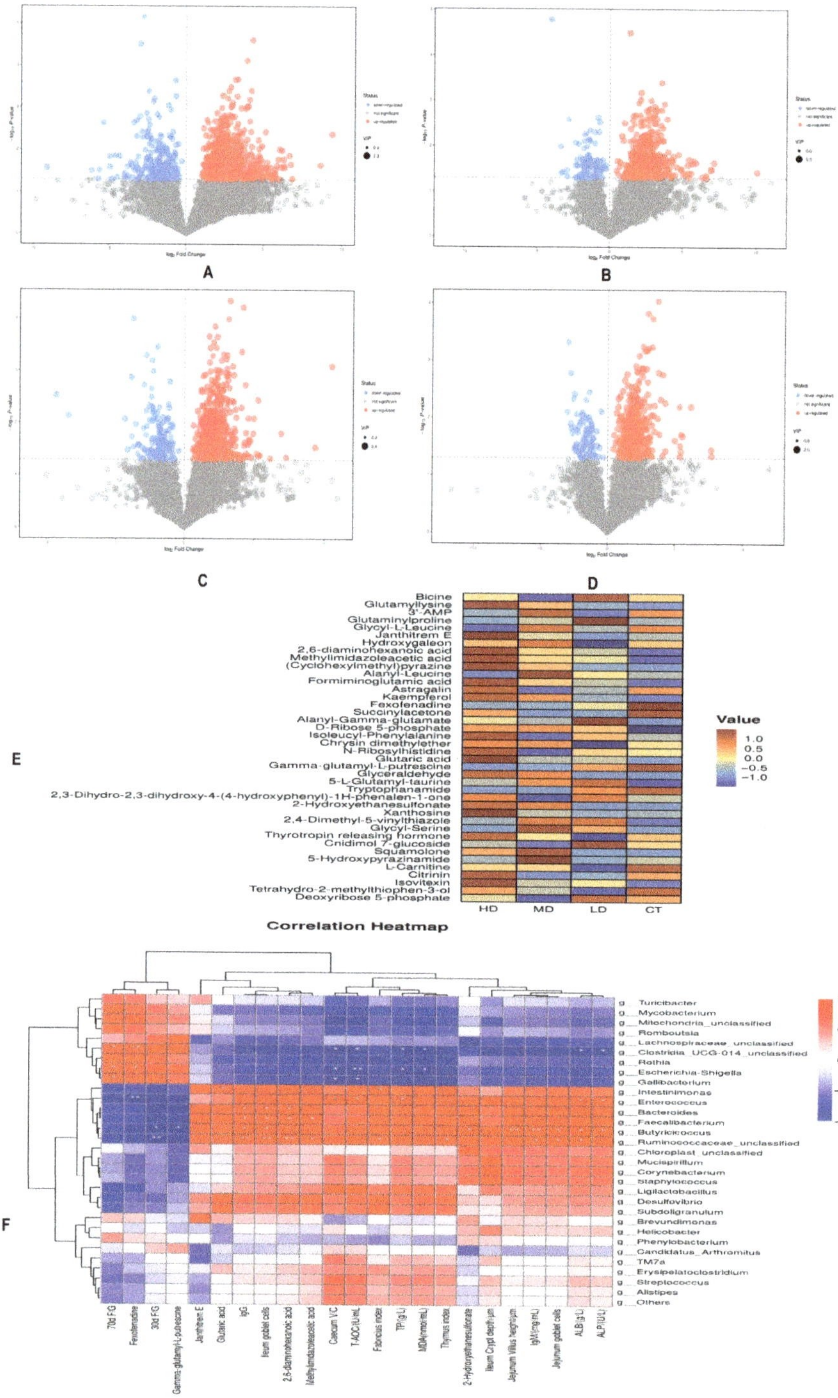

Figure 3. Effects of LP81-FF on the metabolites of the duodenum in Muscovy ducks and Pearson correlations. (**A–D**) Differentially expressed metabolites. (**E**) Forty metabolites with significant

differences in abundance in the duodenum. (**F**) Pearson's correlation cluster heatmap illustrating the relationships among LP81-fermented feed, gut microbiota, derived metabolites, growth performance, and gut morphology of Muscovy ducks under LP81-fermented feed intervention. Darker red and darker blue indicate higher levels of positive and negative correlations, respectively. Significant correlations are marked with asterisks: * $0.01 < p \leq 0.05$; ** $0.001 < p \leq 0.01$.

4. Discussion

As an innovative biological feed resource, microbial-fermented feed demonstrates significant potential for improving poultry production efficiency through enhanced weight gain and optimized feed conversion ratios (FCRs) [20,21]. While previous studies in broilers and Cherry Valley ducks reported increased feed intake alongside growth improvements [22], our findings revealed distinct patterns in Muscovy ducks. LP81-FF supplementation over 35 days significantly reduced average daily feed intake (ADFI) and FCR without compromising weight gain, with these effects persisting through the 70-day trial. Notably, higher LP-81 inclusion levels correlated with progressive reductions in ADFI and FCR. This phenomenon may be attributed to two mechanisms: (1) enzymatic degradation of feed components (crude protein, fiber, and neutral detergent fiber) during fermentation, enhancing nutrient digestibility; *Lactobacillus*-fermented feed improves nutrient decomposition by breaking down complex macromolecules into bioavailable forms. During fermentation, proteolytic activity increases small peptide concentrations, which are more readily absorbed by the intestinal epithelium compared to intact proteins. Simultaneously, carbohydrate fermentation generates short-chain fatty acids (SCFAs), such as acetate and butyrate, serving as energy substrates for enterocytes and enhancing gut barrier function. These metabolic shifts optimize nutrient utilization, reduce undigested feed residues, and alleviate intestinal burdens in ducks. Studies indicate that such feed modifications elevate villus height and digestive enzyme activity, further promoting absorption efficiency. By enhancing gut health and nutrient bioavailability, *Lactobacillus* fermentation not only supports duck growth performance but also aligns with sustainable farming practices by minimizing feed waste and environmental impact [23].

LP81-FF enhances nutrient bioavailability through microbial biotransformation; the reason may be fermentation-liberated amino acids that upregulate jejunal SLC1A1/EAAT3 transporters, accelerating epithelial uptake of both dietary and microbially synthesized nutrients [24]. Concurrently, microbial phytase and xylanase activities degrade antinutritional factors (phytates, β-glucans), releasing encapsulated minerals and starch–protein complexes for absorption, thereby reducing metabolic "waste" in undigested feed—a key factor driving FCR reduction [25]. The water content of fermented feed induces gastric stretch receptor signaling, triggering satiety via CCK-PYY axis activation [26], which lowers dry matter intake without compromising nutrient sufficiency due to enhanced pre-absorptive hydrolysis of macronutrients. This dual optimization—increased nutrient density per gram of ingested dry matter and reduced feed bulk—directly lowers feed procurement and waste management costs by 12–18% in intensive systems.

LP81-FF significantly enhances nutrient metabolism efficiency in ducks through Lactobacillus-fermented pretreatment. Serum biochemical analyses indicate that complex polysaccharides in feed are pre-digested by Lactobacillus LP81 into absorbable monosaccharides, elevating serum glucose levels while reducing reliance on lipid catabolism. The observed glucose elevation, coupled with decreased triglycerides and cholesterol, suggests optimized energy partitioning, likely mediated by microbial-derived short-chain fatty acids (SCFAs): propionate directly inhibits hepatic HMG-CoA reductase activity via allosteric modulation [27], while butyrate activates PPAR-α in hepatocytes, promoting mitochondrial β-oxidation and reducing VLDL secretion [28]. Reduced urea nitrogen levels in the LP81-FF group reflect improved protein utilization efficiency, primarily attributed to *Lactobacillus*-mediated degradation of soybean meal-derived trypsin inhibitors and phytates, thereby enhancing ileal amino acid absorption [29]. Furthermore, gut microbiota-synthesized essential amino acids (e.g., lysine, methionine) compensate for 15–20% of dietary requirements, reducing hepatic deamination and nitrogen waste [30].

Improved antioxidant markers and reduced pro-inflammatory cytokines in LP81-FF-treated ducks highlight systemic metabolic health enhancement. Mechanisms include (1) fermentation-liberated phenolic acids directly scavenging reactive oxygen species (ROS) via electron transfer [31], and (2) butyrate-induced Nrf2 nuclear translocation upregulating phase II detoxification enzymes while suppressing NF-κB-dependent inflammation. These effects redirect ATP toward anabolic processes rather than oxidative stress mitigation [32,33]. LP81-FF significantly increases serum total protein levels, indicating enhanced hepatic and muscular protein synthesis and reduced proteolysis. Dose-dependent elevation of alkaline phosphatase (ALP) activity suggests improved lipid metabolism and growth regulation through (1) enhanced intestinal phosphate absorption for ATP synthesis, (2) hydrolysis of lipoprotein-associated phospholipids to release free fatty acids (FFAs) for β-oxidation, and (3) collagen crosslinking-driven bone mineralization [34]. These metabolic adaptations collectively reduce the feed conversion ratio (FCR), redirecting conserved energy toward muscle deposition rather than inflammatory or oxidative "overhead."

The immunomodulatory effects of LP81-FF are mediated through fermentation-derived bioactive components that differentially regulate systemic versus mucosal antibody production. The significant elevation in serum IgG and IgM levels aligns with the capacity of *Lactobacillus* fermentation to generate conjugated linoleic acids (CLAs) and short-chain fatty acids (SCFAs) [35], which directly modulate B-cell function. Butyrate, a predominant SCFA, enhances IgG synthesis via dual mechanisms: (1) binding to GPR41 receptors on B lymphocytes to activate mTORC1 signaling, driving plasma cell differentiation, and (2) inhibiting histone deacetylases (HDACs), thereby derepressing antibody gene loci through chromatin remodeling. Concurrently, CLAs act as PPAR-γ ligands, amplifying IL-6-dependent STAT3 phosphorylation in B cells, which is critical for IgG class switching [36]. These processes are further potentiated by fermentation-liberated microbial components—peptidoglycan and lipoteichoic acid—that engage TLR2 on dendritic cells, triggering IL-10 production to sustain germinal center reactions in spleen and lymph nodes. The stark contrast between robust systemic IgG/IgM responses and unaltered IgA levels, however, points to strain-specific immunomodulatory properties. LP81 may lack the surface adhesins required for M-cell targeting in Peyer's patches, thereby limiting direct mucosal B-cell priming. Additionally, fermentation-generated high-molecular-weight polysaccharides in LP81-FF may preferentially enter portal circulation via paracellular uptake, bypassing intestinal lymphoid tissue and instead stimulating hepatic B-1 cells—a major source of natural IgM—while failing to activate lamina propria IgA+ plasmablasts. This compartmentalization is functionally significant: elevated IgG enhances Fcγ receptor-mediated phagocytosis of bloodborne pathogens, while IgM provides immediate protection against systemic bacterial invasion through complement activation. The dissociation from prior studies showing fermented feed-induced IgA elevation may reflect strain-dependent variations in postbiotic profiles. For instance, LP81 fermentation might produce lower levels of indole-3-lactic acid—a tryptophan metabolite shown to promote IgA+ B-cell homing to the gut—compared to other *Lactobacillus* strains. Practical implications arise from this specificity: LP81-FF could be strategically combined with mucosally targeted probiotics to achieve comprehensive immune coverage, leveraging its systemic antibody-boosting effects while compensating for localized IgA gaps [37]. Future research should delineate strain-specific epitopes governing immune compartmentalization and quantify the temporal dynamics of antibody isotype switching during prolonged LP81-FF administration.

LP81-FF preserved the intestinal villus architecture—a critical determinant of nutrient absorption efficiency—by sustaining luminal nutrient bioavailability despite reduced feed intake, thereby mitigating the "intestinal starvation" phenotype observed with conventional diets. This maintenance of villus height and crypt depth ratio was accompanied by dose-dependent increases in jejunal and cecal goblet cell densities, suggesting microbial-driven mucin regulation [38]. Fermentation-generated metabolites, particularly butyrate and lactate, directly nourish enterocytes via monocarboxylate transporter (MCT1)-mediated uptake, maintaining villus metabolic activity while suppressing apoptosis-inducing oxidative stress. Concurrently, Lactobacillus-detained proteases degrade feed-derived lectins and trypsin inhibitors that typically blunt villus growth by inducing epithelial hyperproliferation. The thickened mucus layer, enriched in sialylated and sulfated mucins through microbial mucin cross-feeding, creates a gradient that selectively concentrates luminal nutrients near absorptive surfaces while excluding pathogenic competitors. Enhanced mucin sulfonation in high-dose groups, likely mediated by microbial hydrogen sulfide metabolism, improves mucus rheological properties for sustained barrier function. Crucially, preserved villus integrity optimizes brush border enzyme localization, accelerating terminal nutrient hydrolysis proximal to transporters. This microbial–structural synergy enables paradoxical efficiency—reduced physical feed intake yet enhanced nutrient assimilation—by maximizing absorptive surface utility per gram of digesta. The observed cecal goblet cell hyperplasia specifically enhances post-ileal nutrient salvaging through mucin-bound oligosaccharide recycling, a conserved adaptation to nutrient-dense diets [38]. These coordinated adaptations position LP81-FF as a modulator of gut trophic ecology, where microbial metabolites directly couple mucosal maintenance to nutrient harvesting capacity.

LP81-FF induced profound gut microbiota remodeling through nutrient–microbe crosstalk, with dose-dependent enrichment of Bacteroides—specialized degraders of arabinoxylan and pectin via glycoside hydrolase families GH43/51—driving liberation of oligosaccharides that fuel butyrogenesis by Butyricicoccus through cross-feeding [39]. This syntrophic interaction amplifies butyrate production via butyryl-CoA:acetate CoA-transferase, directly energizing colonocytes through mitochondrial β-oxidation while suppressing Escherichia-Shigella via pH-dependent inhibition of Shiga toxin expression. The elevated Bacteroides populations further secrete secondary bile acids that antagonize pathogen membrane integrity through detergent effects, while their capsular polysaccharides competitively block pathogen adhesion to mucin glycans [40]. Metabolomic shifts paralleled these taxonomic changes: increased microbially conjugated amino acids activated aryl hydrocarbon receptor (AhR) signaling in enterocytes, upregulating tight junction proteins (claudin-2, occludin), whereas Butyricicoccus-derived butyrate enhanced barrier function via HDAC3 inhibition and ZO-1 acetylation. The dose-responsive Bacteroides expansion correlated with elevated propionate/acetate ratios, which synergize with microbially synthesized polyamines to upregulate villus mTORC1 signaling, coupling microbial metabolite flux to intestinal nutrient sensing [41]. This microbiota–metabolite axis not only displaced pathogens via niche exclusion but also optimized luminal nutrient solubilization, as Bacteroides-derived xylanases increased starch–protein matrix disintegration, releasing encapsulated amino acids for absorption. The functional convergence of taxonomic shifts (probiotic amplification), metabolic output, and enzymatic activity (polysaccharide hydrolysis) positions LP81-FF as an ecological engineer of gut ecosystem services, where microbial consortia are functionally "rewired" to maximize both barrier protection and nutrient extraction efficiency.

5. Conclusions

Our findings demonstrate that LP81-FF improves Muscovy duck production through multi-faceted mechanisms: enhanced nutrient bioavailability via feed pretreatment, microbiota-mediated metabolic reprogramming, and systemic improvements in antioxidant capacity and immune function. These results position microbial fermentation as a viable strategy for sustainable poultry production optimization.

Author Contributions: Z.L. and S.P. conceptualized the project and wrote the paper; M.Z. and Z.L. performed experiments; X.Z. and H.W. analyzed the data; Z.L., writing—review and editing; F.L., S.P. and T.S. interpreted results of experiments; Z.L., project administration and drafting the manuscript. All authors have read and agreed to the published version of the manuscript.

Funding: This research was founded by the Fujian Provincial Competitive Public Welfare Project (2023R1076), Key Scientific and Technological Project of the Fujian Academy of Agricultural Sciences (KJZD202404).

Institutional Review Board Statement: The animal care and use protocol was approved by the Institutional Animal Care and Use Committee at the Institute of Animal Husbandry and Veterinary Medicine of the Fujian Academy of Agricultural Sciences (202402FJ026, 12 March 2024).

Informed Consent Statement: Not applicable.

Data Availability Statement: The raw data supporting the conclusions of this article will be made available by the authors on request.

Acknowledgments: We thank all of the members of the laboratory for their support and constructive comments, and all authors included in this section have consented to the acknowledgement.

Conflicts of Interest: Author Xiaodong Zhuang was employed by the company Fujian Zhangzhou Changlong Agriculture and Animal Husbandry Co., Ltd. The remaining authors declare that the research was conducted in the absence of any commercial or financial relationships that could be construed as a potential conflict of interest.

References

1. Yan, J.; Zhou, B.; Xi, Y.; Huan, H.; Li, M.; Yu, J.; Zhu, H.; Dai, Z.; Ying, S.; Zhou, W.; et al. Fermented Feed Regulates Growth Performance and the Cecal Microbiota Community in Geese. *Poult. Sci.* **2019**, *98*, 4673–4684. [CrossRef] [PubMed]
2. Guo, W.; Xu, L.-N.; Guo, X.-J.; Wang, W.; Hao, Q.-H.; Wang, S.-Y.; Zhu, B.-C. The Impacts of Fermented Feed on Laying Performance, Egg Quality, Immune Function, Intestinal Morphology and Microbiota of Laying Hens in the Late Laying Cycle. *Animal* **2022**, *16*, 100676. [CrossRef] [PubMed]
3. Ayana, G.U.; Kamutambuko, R.; Wu, J. Probiotics in Disease Management for Sustainable Poultry Production. *Adv. Gut Microbiome Res.* **2024**, *2024*, 4326438. [CrossRef]
4. Ebeid, T.A.; Al-Homidan, I.H.; Fathi, M.M. Physiological and Immunological Benefits of Probiotics and Their Impacts in Poultry Productivity. *Worlds. Poult. Sci. J.* **2021**, *77*, 883–899. [CrossRef]
5. Chang, Y.; Omer, S.H.S.; Liu, L.Y. Research Advance on Application of Microbial Fermented Fodder in Broilers Production: A Short Review. *Open J. Anim. Sci.* **2022**, *12*, 200–209. [CrossRef]
6. Li, Z.; Zhou, H.; Liu, W.; Wu, H.; Li, C.; Lin, F.; Yan, L.; Huang, C. Beneficial Effects of Duck-Derived Lactic Acid Bacteria on Growth Performance and Meat Quality through Modulation of Gut Histomorphology and Intestinal Microflora in Muscovy Ducks. *Poult. Sci.* **2024**, *103*, 104195. [CrossRef]
7. Saleh, A.A.; Shukry, M.; Farrag, F.; Soliman, M.M.; Abdel-Moneim, A.-M.E. Effect of Feeding Wet Feed or Wet Feed Fermented by *Bacillus licheniformis* on Growth Performance, Histopathology and Growth and Lipid Metabolism Marker Genes in Broiler Chickens. *Animals* **2021**, *11*, 83. [CrossRef]

8. Zhou, M.; Zheng, L.; Geng, T.; Wang, Y.; Peng, M.; Hu, F.; Zhao, J.; Wang, X. Effect of Fermented *Artemisia argyi* on Egg Quality, Nutrition, and Flavor by Gut Bacterial Mediation. *Animals* **2023**, *13*, 3678. [CrossRef]
9. Guo, L.; Lv, J.; Liu, Y.; Ma, H.; Chen, B.; Hao, K.; Feng, J.; Min, Y. Effects of Different Fermented Feeds on Production Performance, Cecal Microorganisms, and Intestinal Immunity of Laying Hens. *Animals* **2021**, *11*, 2799. [CrossRef]
10. Li, Z.; Li, C.; Lin, F.; Yan, L.; Wu, H.; Zhou, H.; Guo, Q.; Lin, B.; Xie, B.; Xu, Y.; et al. Duck Compound Probiotics Fermented Diet Alters the Growth Performance by Shaping the Gut Morphology, Microbiota and Metabolism. *Poult. Sci.* **2024**, *103*, 103647. [CrossRef]
11. Council, N.R. *Nutrient Requirements of Poultry*, 9th ed.; The National Academies Press: Washington, DC, USA, 1994. [CrossRef]
12. *NY/T 816-2021*; Nutrient Requirements of Poultry. Ministry of Agriculture and Rural Development: Beijing, China, 2021.
13. *GB/T 6432-2018*; Determination of Crude Protein in Feed by Kjeldahl Nitrogen Determination Method. National Standards Committee: Beijing, China, 2018.
14. *GB/T 6438-2007*; Determination of Crude Ash in Feed. National Committee for Standardization Administration of the People's Republic of China: Beijing, China, 2007.
15. *GB/T 6436-2018*; Determination of Calcium in Feed. National Standardization Administration of China: Beijing, China, 2018.
16. *GB/T 6437-2018*; Determination of Total Phosphorus in Feed by Spectrophotometric Method. National Standardization Administration of China: Beijing, China, 2018.
17. Wang, H.; Wang, X.; Zhan, Y.; Peng, B.; Wang, W.; Yang, L.; Zhu, Y. Predicting the Metabolizable Energy and Metabolizability of Gross Energy of Conventional Feedstuffs for Muscovy Duck Using in Vitro Digestion Method. *J. Anim. Sci.* **2023**, *101*, skad018. [CrossRef] [PubMed]
18. *GB/T 6435-2006*; Determination of Moisture and Other Volatile Mater Content in Feeds. General Administration of Quality Supervision, Inspection and Quarantine of the People's Republic of China, Standardization Administration of China: Jinan, China, 2006.
19. *GB/T 6434-2022*; Determination of Crude Fiber Content in Feeds. General Administration of Quality Supervision, Inspection and Quarantine of the People's Republic of China, Standardization Administration of China: Jinan, China, 2022.
20. Zhu, X.; Tao, L.; Liu, H.; Yang, G. Effects of Fermented Feed on Growth Performance, Immune Organ Indices, Serum Biochemical Parameters, Cecal Odorous Compound Production, and the Microbiota Community in Broilers. *Poult. Sci.* **2023**, *102*, 102629. [CrossRef] [PubMed]
21. Chen, J.-Y.; Yu, Y.-H. *Bacillus subtilis*-Fermented Products Ameliorate the Growth Performance and Alter Cecal Microbiota Community in Broilers under Lipopolysaccharide Challenge. *Poult. Sci.* **2021**, *100*, 875–886. [CrossRef]
22. Zhang, M.; Yu, A.; Wu, H.; Xiong, X.; Li, J.; Chen, L. *Lactobacillus acidophilus* and *Bacillus subtilis* Significantly Change the Growth Performance, Serum Immunity and Cecal Microbiota of Cherry Valley Ducks during the Fattening Period. *Anim. Sci. J.* **2024**, *95*, e13946. [CrossRef]
23. Oh, S.T.; Zheng, L.; Kwon, H.J.; Choo, Y.K.; Lee, K.W.; Kang, C.W.; An, B.K. Effects of Dietary Fermented *Chlorella vulgaris* (CBT(®)) on Growth Performance, Relative Organ Weights, Cecal Microflora, Tibia Bone Characteristics, and Meat Qualities in Pekin Ducks. *Asian-Australas. J. Anim. Sci.* **2015**, *28*, 95–101. [CrossRef]
24. Ye, J.-L.; Gao, C.-Q.; Li, X.-G.; Jin, C.-L.; Wang, D.; Shu, G.; Wang, W.-C.; Kong, X.-F.; Yao, K.; Yan, H.-C.; et al. EAAT3 Promotes Amino Acid Transport and Proliferation of Porcine Intestinal Epithelial Cells. *Oncotarget* **2016**, *7*, 38681–38692. [CrossRef]
25. Farhat-Khemakhem, A.; Blibech, M.; Boukhris, I.; Makni, M.; Chouayekh, H. Assessment of the Potential of the Multi-Enzyme Producer *Bacillus amyloliquefaciens* US573 as Alternative Feed Additive. *J. Sci. Food Agric.* **2018**, *98*, 1208–1215. [CrossRef]
26. Zanchi, D.; Depoorter, A.; Egloff, L.; Haller, S.; Mählmann, L.; Lang, U.E.; Drewe, J.; Beglinger, C.; Schmidt, A.; Borgwardt, S. The Impact of Gut Hormones on the Neural Circuit of Appetite and Satiety: A Systematic Review. *Neurosci. Biobehav. Rev.* **2017**, *80*, 457–475. [CrossRef]
27. Schönfeld, P.; Wojtczak, L. Short- and Medium-Chain Fatty Acids in Energy Metabolism: The Cellular Perspective. *J. Lipid Res.* **2016**, *57*, 943–954. [CrossRef]
28. Weng, H.; Endo, K.; Li, J.; Kito, N.; Iwai, N. Induction of Peroxisomes by Butyrate-Producing Probiotics. *PLoS ONE* **2015**, *10*, e0117851. [CrossRef]
29. Li-Wen, S.U.; Cheng, Y.H.; Hsiao, S.H.; Han, J.C.; Yu-Hsiang, Y.U. Optimization of Mixed Solid-State Fermentation of Soybean Meal by *Lactobacillus species* and *Clostridium butyricum*. *Pol. J. Microbiol.* **2018**, *67*, 297–305.
30. Erskine, E.; Ozkan, G.; Lu, B.; Capanoglu, E. Effects of Fermentation Process on the Antioxidant Capacity of Fruit Byproducts. *ACS Omega* **2023**, *8*, 4543–4553. [CrossRef] [PubMed]
31. Ayayee, P.A.; Thomas, L.; Cristina, R.; Felton, G.W.; Ferry, J.G.; Kelli, H. Essential Amino Acid Supplementation by Gut Microbes of a Wood-Feeding Cerambycid. *Environ. Entomol.* **2016**, *45*, 66–73. [CrossRef] [PubMed]

32. Mudronová, D.; Karaffová, V.; Koščová, J.; Bartkovský, M.; Marcincáková, D.; Popelka, P.; Klempová, T.; Certík, M.; Macanga, J.; Marcincák, S. Effect of Fungal Gamma-Linolenic Acid and Beta-Carotene Containing Prefermented Feed on Immunity and Gut of Broiler Chicken. *Poult. Sci.* **2018**, *97*, 4211–4218. [CrossRef]
33. Memon, M.A.; Dai, H.; Wang, Y.; Xu, T.; Aabdin, Z.U.; Bilal, M.S.; Chandra, R.A.; Shen, X. Efficacy of Sodium Butyrate in Alleviating Mammary Oxidative Stress Induced by Sub-Acute Ruminal Acidosis in Lactating Goats. *Microb. Pathog.* **2019**, *137*, 103781. [CrossRef]
34. Yang, Y.; Rader, E.; Peters-Carr, M.; Bent, R.C.; Smilowitz, J.T.; Guillemin, K.; Rader, B. Ontogeny of Alkaline Phosphatase Activity in Infant Intestines and Breast Milk. *BMC Pediatr.* **2019**, *19*, 2. [CrossRef]
35. Wang, E.; Cha, M.; Wang, S.; Wang, Q.; Wang, Y.; Li, S.; Wang, W. Feeding Corn Silage or Grass Hay as Sole Dietary Forage Sources: Overall Mechanism of Forages Regulating Health-Promoting Fatty Acid Status in Milk of Dairy Cows. *Foods* **2023**, *12*, 303. [CrossRef]
36. Sanchez, H.N.; Shen, T.; Taylor, J.R.; Munoz, K.; Zan, H.; Casali, P. Short-Chain Fatty Acid (SCFA) Histone Deacetylase Inhibitors Produced by Gut Microbiota Regulate Selected MicroRNAs to Modulate Local and Systemic Antibody Responses. *J. Immunol.* **2016**, *196*, 127.6. [CrossRef]
37. Jiang, D.; Yang, M.; Xu, J.; Deng, L.; Hu, C.; Zhang, L.; Sun, Y.; Jiang, J.; Lu, L. Three-Stage Fermentation of the Feed and the Application on Weaned Piglets. *Front. Vet. Sci.* **2023**, *10*, 1123563. [CrossRef]
38. López-García, Y.R.; Gómez-Rosales, S.; Angeles, M.d.L.; Jiménez-Severiano, H.; Merino-Guzman, R.; Téllez-Isaias, G. Effect of the Addition of Humic Substances on Morphometric Analysis and Number of Goblet Cells in the Intestinal Mucosa of Broiler Chickens. *Animals* **2023**, *13*, 212. [CrossRef]
39. Vasaï, F.; Ricaud, K.B.; Cauquil, L.; Daniel, P.; Peillod, C.; Gontier, K.; Tizaoui, A.; Bouchez, O.; Combes, S.; Davail, S. *Lactobacillus sakei* Modulates Mule Duck Microbiota in Ileum and Ceca during Overfeeding. *Poult. Sci.* **2014**, *93*, 916–925. [CrossRef] [PubMed]
40. Zhang, F.; Yang, J.; Zhan, Q.; Shi, H.; Li, Y.; Li, D.; Li, Y.; Yang, X. Dietary Oregano Aqueous Extract Improves Growth Performance and Intestinal Health of Broilers through Modulating Gut Microbial Compositions. *J. Anim. Sci. Biotechnol.* **2023**, *14*, 77. [CrossRef] [PubMed]
41. Reges, B.M.; da Silva Oliveira, F.A.; Fonteles, T.V.; Rodrigues, S. Changes in Human Colonic Microbiota Promoted by Synbiotic Aai Juice Composed of Gluco-Oligosaccharides, Dextran, and *Bifidobacterium breve* NRRL B-41408. *Foods* **2024**, *13*, 4121. [CrossRef] [PubMed]

fermentation

MDPI

Article

Fermented Chive (*Allium schoenoprasum*) with *Lactobacillus plantarum*: A Potential Antibiotic Alternative Feed Additive for Broilers Challenged with *Escherichia coli*

Phan Vu Hai [1,*], Le Xuan Anh [1,2] and Nguyen Xuan Hoa [1]

[1] Faculty of Animal Sciences and Veterinary Medicine, University of Agriculture and Forestry, Hue University, Hue City 530000, Vietnam; lexuananh@huaf.edu.vn (L.X.A.); nguyenxuanhoa@huaf.edu.vn (N.X.H.)
[2] Department of Veterinary Medicine, College of Veterinary Medicine, National Pingtung University of Science and Technology, Pingtung 91201, Taiwan
* Correspondence: phanvuhai@hueuni.edu.vn

Abstract: This study aimed to evaluate the effects of fermented chive (*Allium schoenoprasum*) with *Lactobacillus plantarum* 1582 (FC) as an alternative to antibiotics for controlling *Escherichia coli* infection in broiler chickens. A total of 250 J-Dabaco male chickens were allocated into five experimental groups: NC (negative control), PC (positive control), FC1 (1% FC), FC3 (3% FC), and AB (antibiotic treatment). The PC, FC1, FC3, and AB groups were challenged with *E. coli* ExPEC_A338 on day 8 and monitored until day 35. The results indicated that FC supplementation, particularly at 3% (FC3 group), significantly improved body weight gain, feed intake, the survival rate, and the production efficiency index (PEI). The FC3 group exhibited optimal performance, potentially due to enhanced immune responses, as evidenced by higher IgA and IgG levels, and favorable cytokine regulation. Additionally, FC maintained intestinal epithelial integrity by upregulating tight junction proteins (ZO-1, Claudin-2) and reducing inflammatory responses (IFN-γ, TNF-α). Furthermore, FC3 demonstrated the ability to inhibit pathogenic bacteria (*Salmonella* spp., *E. coli*), promote beneficial *Lactobacillus* spp., and enhance intestinal mucosal morphology (villus height and crypt depth). These findings suggest that FC supplementation, particularly at 3%, is a promising natural alternative to antibiotics for controlling *E. coli* infections in broiler production.

Keywords: *Allium schoenoprasum*; natural alternative to antibiotics; broilers; *Escherichia coli*; *Lactobacillus plantarum*

1. Introduction

Escherichia coli (*E. coli*) is a common bacterial pathogen in poultry worldwide, causing severe health issues and reduced productivity [1]. *E. coli* infections in poultry can lead to diseases, such as necrotic enteritis, airsacculitis, and septicemia, resulting in significant economic losses in the poultry industry [2]. Moreover, infected poultry and contaminated poultry products can serve as sources of zoonotic transmission, posing a public health risk. Therefore, controlling the spread of *E. coli* in poultry farming is of critical importance.

Antibiotics have been widely used for the prevention and treatment of *E. coli* infections in poultry. However, antibiotic overuse has contributed to the rise of antimicrobial resistance (AMR), posing a serious threat to both animal and human health. Additionally, antibiotic use can disrupt gut microbiota balance, negatively impacting poultry health and productivity. As a result, there is an urgent need to develop effective, safe, and environmentally friendly antibiotic alternatives.

Recently, medicinal plants have gained attention as potential alternatives due to their bioactive compounds with antimicrobial, antioxidant, and immunomodulatory properties. Among them, chive (*Allium schoenoprasum*), a widely used spice in Vietnam, is rich in flavonoids (especially quercetin) and organosulfur compounds, which may improve animal health [3]. However, direct dietary supplementation of chive may be limited in efficacy because plant cell walls restrict the bioavailability of these compounds [4].

Fermentation is an effective approach to enhance bioactivity by increasing the release of polyphenols, free amino acids, and organosulfur compounds while also boosting antioxidant properties [5]. Furthermore, fermentation supports the proliferation of beneficial gut bacteria, contributing to gut microbiota balance and intestinal health [6]. However, spontaneous fermentation using native microorganisms may be difficult to control and could lead to the production of harmful substances [7]. Therefore, selecting an appropriate bacterial strain is crucial for optimizing chive fermentation and maximizing its benefits.

Lactobacillus plantarum is one of the most commonly used probiotic species, known for its ability to adhere to the intestinal mucosa, produce digestive enzymes, modulate gut microbiota, and enhance immune responses [8]. Despite these benefits, *L. plantarum* alone may have limited efficacy in disease prevention. Combining probiotics with fermented medicinal plants is considered a promising strategy to enhance disease resistance and promote poultry health.

Our recent study successfully isolated *L. plantarum* 1582 from free-range native chicken feces, noting its capacity for chive fermentation, *E. coli* inhibition, and survivability in the chicken digestive tract [9]. Based on the potential of this strain, the present study aims to evaluate the effects of fermented chive with *L. plantarum* 1582 on the growth performance, gut health, and immune responses of *E. coli*-challenged broiler chickens.

2. Materials and Methods

2.1. Preparation of Fermented Chive

Chive (*Allium schoenoprasum*; GenBank ID: NC_057575.1), aged 4–5 months and cultivated under VietGAP biosecurity standards (TCVN 11892-1:2017) [10], was used as the raw material for the preparation. After preliminary processing, the chive was finely ground for fermentation.

Fermentation was conducted following the method described by Hai et al. (2024) [9], with some modifications. Specifically, the chive bulbs were fermented with *Lactobacillus plantarum* 1582 (GenBank ID: MT597487.1) at a concentration of 10^8 CFU/mL in a medium containing 5% NaCl and 3% glucose. The mixture was incubated at 37 °C under anaerobic conditions with shaking at 60 rpm for 72 h.

Following fermentation, both the fermented and non-fermented preparations were mixed with starch at a 3:7 ratio, dried at 50 °C to a moisture content of 3%, and finely ground. The final fermented product contained *Lactobacillus* spp. at a density of approximately $2.7–3 \times 10^8$ CFU/g and was analyzed for key bioactive compounds (Table 1).

2.2. Pathogenic Bacteria

The virulent *Escherichia coli* strain ExPEC_A338 (GenBank ID: CP142559.1) was isolated from the feces of indigenous chickens suffering from diarrhea suspected to be caused by *E. coli*. The strain was preserved at the Bacteriology and Infectious Diseases Laboratory, Faculty of Animal Science and Veterinary Medicine, Hue University of Agriculture and Forestry. Prior to experimentation, the bacteria were activated in Luria–Bertani broth for three generations, centrifuged (3500× *g*, 15 min, 4 °C), and washed twice with phosphate-buffered saline (PBS, pH 7.2).

Table 1. Bioactive and nutritional compounds in fresh chive and chive fermented with *L. plantarum*.

Component/Compound	Fresh Chive	Fermented Chive	Determination Method
Polyphenol (mg/g)	10–15	15–20	Folin-Ciocalteu
Quercetin (mg/g)	2–4	3–6	UV-Vis
Sulfur compounds			
Allicin (mg/kg)	1–1.5	1–1.5	GC-MS
Thiosulfate (mg/g)	5–7	~2–3	GC
S-allyl cysteine (mg/g)	0.1–0.3	1–3	GC-MS
Organic acids			
Lactic acid (%)	negligibility	0.5–1.5	HPLC
Acetic acid (%)	negligibility	0.1–0.5	HPLC
Citric acid (%)	~0.2–0.5	~0.5–1	HPLC
Crude protein (g/kg)	16.3–20.1	16.2–22.3	AOAC
Crude fat (g/kg)	6–9	3–4	AOAC
Crude fiber (g/kg)	22.3–26.8	21.4–24.5	AOAC
Metabolizable energy (kCal/100g)	330–370	360–380	AOAC

2.3. Experimental Animals

Male J-Dabaco broiler chickens were obtained from Dabaco Company (Vietnam) and reared for the entire experimental period (1–35 days of age) in a well-maintained cross-ventilated housing system. During the first week, the chicks were brooded together in a circular brooder with microbiologically treated rice husk bedding. From days 8 to 35, the birds were housed in wire cages (dimensions: 0.9 × 0.5 × 0.5 m). Continuous fluorescent lighting was provided throughout the experiment. The ambient temperature was maintained at 35 °C during the first two weeks and then gradually reduced to 25 °C until the end of the trial. The diet was formulated using locally sourced ingredients, such as yellow corn, soybean meal, and fishmeal, to meet the nutritional standards for broiler chickens, as specified by the Vietnam Ministry of Agriculture and Rural Development (10 TCN 661-2005) [11]. The birds had ad libitum access to feed and antibiotic-free drinking water throughout the experiment.

The raw materials and medicinal ingredients (if applicable) were mixed using an HM-150 mixer (Hai Minh Co., Ltd., HCM City, Vietnam; 5.5 kW, 50 rpm, ~1500 kg/h). The mixed feed was then pelleted using an S270 pellet mill (Binh Quan Group, HCM city, Vietnam; 11 kW, 1450 rpm, ~500 kg/h) with a die size of 3 mm for the starter phase and 5 mm for the finisher phase, maintaining a moisture content of approximately 3%.

2.4. Experimental Design

A total of 250 one-day-old male chicks were randomly assigned to five groups, with five replicates per group (10 birds per cage): (1) negative control (NC, no supplementation, no *E. coli* infection), (2) positive control (PC, no supplementation, *E. coli* infection), (3) low-dose of FC (1%, FC1), (4) high-dose of FC (3%, FC3), and (5) antibiotic-treated group with Terra-Neocine (AB). Table 2 presents the ingredients and the calculated nutritional composition of the experimental diets. The chickens in groups 2–5 were orally challenged with *E. coli* ExPEC_A338 (2.9 × 10^8 CFU/mL) on day 8. The chickens in group 5 were supplemented with Terra-Neocine (Mebipha Co., Hanoi, Vietnam) at a dose of 2 g/1 L of drinking water, continuously used for 3 days before *E. coli* infection and 5 days after. The supplemented treatments constituted 1% and 3% of the diet throughout this study.

Table 2. Ingredients and the calculated composition of the experimental diets (as-fed basis, %, unless otherwise stated).

Nutrient	Starter Phase (<19 Days of Age)					Finisher Phase (≥19 Days of Age)				
	NC	PC	AB	FC1	FC3	NC	PC	AB	FC1	FC3
Ingredient composition										
Fermented chive (FC)		0		1	3		0		1	3
Yellow corn		47		47.1	47.2		58.5		58.6	58.6
Soybean meal (36.7% CP)		44.28		43.08	40.98		33.78		32.58	30.68
Fish meal		5		5.1	5.1		4		4.1	4
$CaCO_3$ (38%)		2		2	2		2		2	2
$CaHPO_4$		1		1	1		1		1	1
Sodium chloride		0.3		0.3	0.3		0.3		0.3	0.3
Choline chloride (50%)		0.02		0.02	0.02		0.02		0.02	0.02
DL-Methionine (99.5%)		0.2		0.2	0.2		0.2		0.2	0.2
Vitamin premix [1]		0.1		0.1	0.1		0.1		0.1	0.1
Mineral premix [2]		0.1		0.1	0.1		0.1		0.1	0.1
Calculated nutritional values										
Crude protein		22.2		21.3	22.2		18.0		18.1	18.1
Crude fat		4.48		4.43	4.32		5.48		5.40	5.36
Crude fiber		5.05		5.06	5.13		5.05		5.08	5.09
Calcium		1.10		1.09	1.06		1.00		1.03	1.01
Phosphorus		0.51		0.50	0.50		0.44		0.45	0.45
Lysine		1.32		1.30	1.30		1.04		1.02	1.01
Metabolizable energy (kCal/kg)		3000		3000	3000		3200		3200	3200

[1] Each kilogram of vitamin premix contained 10 mg nicotinamide, 0.02 mg cholecalciferol, 0.3 mg folic acid, 2 mg pyridoxine HCl, 1.8 mg all-trans-retinyl acetate, 8 mg cyanocobalamin, 2.2 mg menadione, 8.3 mg alpha-tocopherol acetate, 160 mg choline chloride, and 20 mg D-biotin. [2] Each kilogram of mineral premix contained 60 µg selenium (Se), 200 µg cobalt (Co) from $CoSO_4$, 800 µg iodine (I) from KI, 2 mg copper (Cu) from $CuSO_4·5H_2O$, 24 mg zinc (Zn) from ZnO, 16 mg iron (Fe) from $FeSO_4·7H_2O$, and 32 mg manganese (Mn) from $MnSO_4·H_2O$.

On day 11 of the experiment, three birds per replicate were randomly selected for sample collection. Blood samples were collected to assess serum immunoglobulin levels, while small intestinal samples were obtained for immune gene expression analysis, and ileal samples were collected to evaluate microbial density. At the end of the experiment (day 35), three chickens per replicate were again randomly selected for sample collection to examine immune organs and perform histopathological analysis (Figure 1).

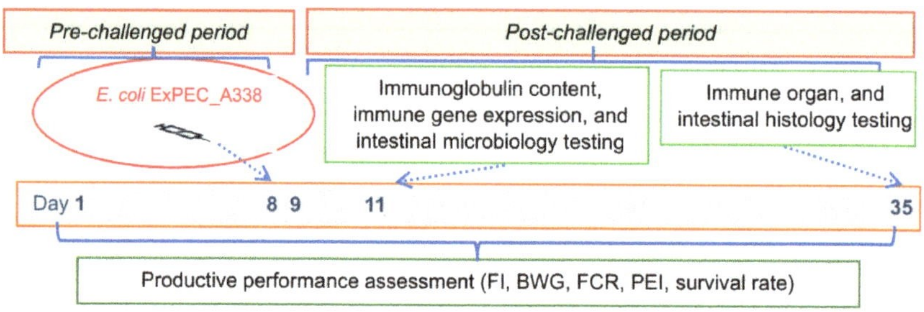

Figure 1. Flowchart of the evaluation of parameters over time.

2.5. Indicators and Research Methods

Growth performance assessment: The feed intake (FI) of the chicks in each treatment was determined daily at 07:00 h, as well as the survival rate and body weight (BW), to determine the following indicators: Weight gain (BWG) = Final BW − Initial BW; Feed conversion ratio (FCR) = FI/BWG; and Production efficiency index (PEI) = BWG (kg) × Survival rate (%)/Number of experimental days (day) × FCR × 100 [12]

Immunoglobulin content: Blood samples were centrifuged (3000 rpm, 10 min), and then, the serum was separated and stored at -20 °C. Concentrations of IgA (MBS705241), IgM (MBS706158), and IgG (MBS260043) were determined by ELISA (MyBioSource, San Diego, CA, USA), optical density was measured at 450 nm, and immunoglobulin concentrations were calculated based on the standard curve.

Immune gene expression: Total RNA was extracted from the ileal mucosa using a Treezol kit (Invitrogen, Carlsbad, CA, USA) and checked using a Nanodrop Lite (Thermo Fisher, Waltham, MA, USA). cDNA was synthesized using a FIRE Script RT cDNA kit (Solid Biodyne, Tartu, Estonia). Immune genes (IL-4, IL-1β, TNF-α, IFN-γ, ZO-01, claudin-2, occludin) and reference gene GAPDH were synthesized by Sangon Biotech Co., Ltd. (Shanghai, China) (Table 3). RT-qPCR was performed on a Quant Studio™ 5 (Thermo Fisher, USA) with an amplification cycle consisting of 95 °C (1 min), 40 cycles of 95 °C (15 s), and 60 °C (60 s). Gene expression was calculated using the 2−ΔΔCt method (Livak and Schmittgen, 2001) [13] with GAPDH as the reference gene.

Table 3. Primer sequences used in RT-qPCR.

Gene	Sequence Primer (5′ to 3′)			Genbank ID
	Forward Primer	Reverse Primer	Size (bp)	
Tight-binding protein				
ZO-1	CTTCAGGTGTTTCTCTTCCTCCTC	CTGTGGTTTCATGGCTGGAT	121	XM_413773.4
Occludin	GCAGATGTCCAGCGGTFC1CFC1C	CGAAGAAGCAGATGAGGCAGAG	89	NM_205128.1
Claudin-2	CAFC1CTCCTGGGTCTGGTTGGT	GACAGCCATCCGCATCTTCT	198	NM_001013611.2
Pro-inflammatory cytokines				
IL-4	GTGCCCACGCT GTGCTFC1C	AGGAAACCTCT CCCTGGATGTC	82	GU119892.1
IL-1β	GFC1CCGAGFC1C AACCCCTGC	AGCAACGGGAC GGT AATGAA	204	NM_204524.1
TNF-α	CTCAGGACAGC CFC1TGCCAACA	CCACCACACGA CAG CCAAGT	177	XM_015294125.2
IFN-γ	CCTCGCAACCT TCACCTCAC	CGCTGFC1ATCG TTG TCTTGGAG	76	FJ977575.1
GAPDH	AACTTTGGCAT TGTGGAGGG	ACGCTGGGATG ATGTTCTGG	130	NM_204305.1

Microbial density determination: *Escherichia coli*, *Salmonella* spp., and *Lactobacillus* spp. were isolated from ileal digesta using selective media, including eosin methylene blue (EMB) agar (supplemented with 1% $CaCO_3$), *Salmonella–Shigella* (SS) agar, and de Man, Rogosa, and Sharpe (MRS) agar, following ISO standards 13349/2001, 6579/2003, 7937/2004, and ISO/Dis 11290/1994. A 1 g sample was homogenized with 9 mL of sterile 0.1% peptone solution and serially diluted (10^{-1} to 10^{-7}). Three dilutions (10^{-3}, 10^{-5}, 10^{-7}) were selected for plating, with 100 μL spread onto appropriate selective media in Petri dishes. The plates were incubated at 37 °C for 24 h, and colony-forming units (CFUs) were counted and expressed as $\log_{10}$ (CFU/g).

Immune organ index: The immune organs, including the bursa of Fabricius, thymus, and spleen, were weighed. The immune organ index was calculated as the organ weight (g) per 100 g of live body weight.

Histopathological examination: Duodenum, jejunum, and ileal samples were fixed in 10% formalin saline for 24 h and processed according to the method described by Layton et al. (2019) [14]. Tissue segments (1 cm) were dehydrated in graded ethanol solutions (70%,

90%, 100%) for 2 h each, cleared with xylene (twice for 1 h each), and embedded in molten paraffin wax (twice for 1 h each) before block embedding. Tissue sections (5 μm thick) were prepared using a rotary microtome (Leica RT 25, New Milton, Hampshire, UK), mounted on glass slides, dried, and stained with hematoxylin and eosin (H&E). Morphological characteristics, including villus height and crypt depth, were analyzed using a Nikon Eclipse E200 microscope (Nikon Instruments Inc., Melville, NY, USA) and Image-Pro Plus software (version 6.0, Media Cybernetics, Rockville, MD, USA).

2.6. Statistical Analysis

The data collected in this study were processed using Excel 2019 and statistically analyzed using IBM. SPSS 22.0. The results were expressed as mean ± standard error of the mean (SEM). A one-way ANOVA followed by post hoc Bonferroni tests was used to determine statistical significance, while the chi-square (χ^2) test was employed to analyze percentage differences at a significance level of $\alpha = 0.05$.

3. Results

3.1. Effect of Supplementary Fermented Chive on the Productivity of Broiler Chickens

The results presented in Table 4 and Figure 2 indicate that during the initial phase of the experiment (prior to *E. coli* infection), there were no significant differences ($p > 0.05$) among the groups in terms of body weight gain (BWG), feed intake (FI), or the feed conversion ratio (FCR). All groups maintained a 100% survival rate, suggesting that dietary supplementation with FC-based products (FC3 or FC1) did not adversely affect the normal growth of the chickens.

Table 4. Effect of fermented chive supplementation on the growth performance of broiler chickens.

Target	NC	PC	FC1	FC3	AB	Pooled SEM	*p*-Value
			Pre-challenged period (1–7 days old)				
BWG (g/head)	25.5	26.9	26.5	25.6	26.6	0.25	0.195
FI (g/head)	45.5	45.5	45.5	45.6	45.6	0.29	0.341
FCR	1.78	1.74	1.76	1.81	1.75	0.16	0.166
			Post-challenged period (8–35 days old)				
BWG (g/head)	699.3 ^{a}	633.7 ^{b}	686.7 ^{ab}	714.7 ^{a}	699.4 ^{a}	33.29	<0.001
FI (g/head)	1588 ^{b}	1471.3 ^{c}	1623.2 ^{ab}	1668.2 ^{a}	1630.8 ^{ab}	63.52	<0.001
FCR	2.28	2.32	2.35	2.33	2.33	0.13	0.069
			Full time (1–35 days old)				
BWG (g/head)	724.4 ^{ab}	659.4 ^{c}	711.9 ^{b}	739.5 ^{a}	725.0 ^{ab}	34.92	<0.001
FI (g/head)	1634.4 ^{ab}	1516.5 ^{c}	1668.4 ^{ab}	1713.5 ^{a}	1576.2 ^{bc}	89.51	<0.001
FCR	2.26 ^{ab}	2.30 ^{ab}	2.34 ^{a}	2.32 ^{a}	2.17 ^{b}	0.131	<0.001
PEI	89.94 ^{a}	57.37 ^{c}	72.92 ^{b}	83.91 ^{a}	85.77 ^{a}	3.57	<0.001

Note: In the same row, values with different lower-case letters (a, b and c) indicate statistically significant differences ($p < 0.05$). Weight gain (BWG), feed intake (FI), feed conversion ratio (FCR), production efficiency index (PEI).

Following *E. coli* challenge, the FC3 group exhibited the highest BWG (714.6 g/bird), which was significantly higher than that of the PC group ($p = 0.025$) and comparable to that of the AB group. The feed intake in the FC3 group was also the highest (1668.1 g/bird), suggesting improved palatability or increased metabolic demand, which contributed to greater weight gain. Overall, both the FC3 and FC1 groups mitigated the adverse effects of E. coli infection on growth performance, with improvements in BWG of 11.2% and 10.8%

and increases in FI of 11.3% and 11.0% compared to the PC group. Notably, the FC3 group showed a statistically significant difference ($p < 0.05$) in comparison to the PC group.

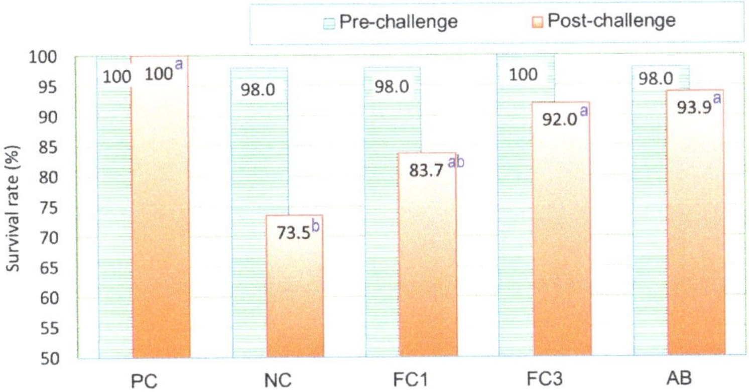

Figure 2. Survival rate of chickens before and after *E. coli* challenge. Values with different lower-case letters (a,b) indicate statistically significant differences ($p < 0.05$).

Regarding survival rate, the FC3 group achieved 93.34%, which was significantly higher than the PC group (73.34%) and comparable to the AB group (96.68%). The chickens challenged with *E. coli* exhibited hemorrhagic intestines, purulent pericarditis, caseous abdominal exudate, and fibrinous liver deposits (Figure 3). The production efficiency index (PEI) of the FC3 group (83.91) was also significantly higher ($p < 0.05$) than that of the FC1 (72.92) and PC (57.37) groups and comparable to the AB (85.77) and NC (89.94) groups.

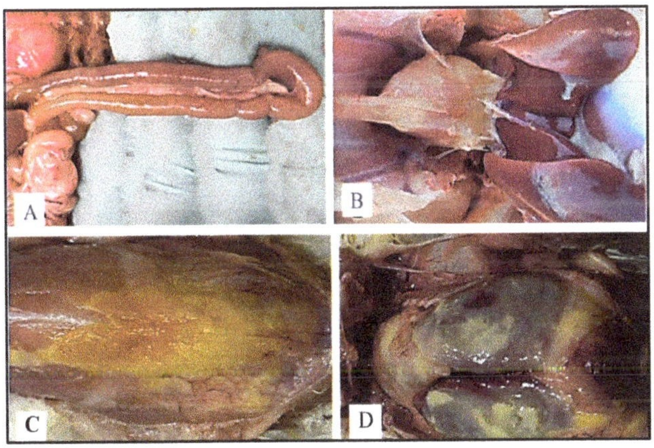

Figure 3. Gross pathological lesions in chickens following *E. coli* challenge. (**A**) Hemorrhagic duodenal mucosa, intestinal congestion, and hemorrhage. (**B**) Thickened pericardium with purulent exudate and the presence of purulent fluid in the pericardial cavity. (**C**) Yellow, cheese-like exudate oozing from the abdominal cavity. (**D**) Grayish-white fibrinous exudate and yellow caseous deposits covering the liver surface.

3.2. *Effect of Fermented Chive Supplementation on Immune Function*

The results presented in Table 5 indicate that the bursa fabricius index (2.44 g/kg) and spleen index (2.25 g/kg) in the FC3 group were significantly higher ($p < 0.05$) compared to the other groups. Meanwhile, the thymus index showed no significant differences among the experimental groups, ranging from 3.4 g/kg to 3.9 g/kg.

Table 5. Immune organ indices and serum immunoglobulin levels in experimental broiler chickens.

Treatment	Immune Organ Index (g/kg BW)			Immunoglobulin Concentration (g/L)		
	Bursa of Fabricius	Spleen	Thymus	IgA	IgM	IgG
NC	2.01 ^{b}	1.82 ^{b}	3.44	3.23 ^{b}	3.22 ^{b}	1.93
PC	1.98 ^{b}	1.79 ^{b}	3.82	2.19 ^{a}	2.18 ^{a}	1.88
FC1	2.04 ^{b}	1.83 ^{b}	3.74	3.22 ^{b}	3.63 ^{b}	2.02
FC3	2.44 ^{a}	2.25 ^{a}	3.92	2.90 ^{b}	3.32 ^{b}	1.96
AB	1.99 ^{b}	1.79 ^{b}	3.51	2.46 ^{ab}	2.69 ^{ab}	1.98
Pooled SEM	0.02	0.02	0.06	0.04	0.04	0.03
p-value	<0.01	<0.01	0.262	<0.01	<0.01	0.618

Note: In the same row, values with different lower-case letters (a and b) indicate statistically significant differences ($p < 0.05$).

The immunoglobulin (IgA, IgM, IgG) concentrations in the serum of the experimental chickens at 35 days of age (27 days post-infection) revealed significant differences in the IgA and IgM levels among treatments ($p < 0.01$). Specifically, the *E. coli*-challenged group (PC) (2.18–2.19 g/L) exhibited significantly lower values ($p < 0.01$) compared to the NC and FC1 groups (3.22–3.23 and 3.22–3.63 g/L, respectively).

The results in Table 6 show that the FC3 group exhibited a significant improvement in the intestinal epithelial barrier, with ZO-1 levels being 3.7 times higher ($p = 0.025$) than those in the PC group. Claudin-2 expression also increased significantly ($p = 0.019$), whereas occludin levels showed no significant differences among the groups.

Table 6. Effects of fermented chive supplementation on immune gene expression levels in broiler chickens.

Gene	NC	PC	FC1	FC3	AB	Pooled SEM	*p*-Value
			Tight-binding protein				
ZO-1	0.93 ^{b}	0.82 ^{b}	0.95 ^{b}	3.02 ^{a}	0.96 ^{b}	0.13	<0.001
Occludin	1.12	1.16	0.97	1.08	1.19	0.13	0.146
Claudin-2	0.91 ^{bc}	0.42 ^{d}	1.23 ^{b}	2.75 ^{a}	1.15 ^{b}	0.14	<0.001
			Pro-inflammatory cytokines				
IL-4	2.33 ^{b}	1.85 ^{b}	3.12 ^{a}	3.29 ^{a}	3.38^{a}	0.19	<0.001
IL-1β	1.24	1.27	1.37	1.34	1.34	0.24	0.949
TNF-α	1.14 ^{c}	2.01 ^{a}	1.62 ^{b}	1.22 ^{c}	1.19 ^{c}	0.08	<0.001
IFN-γ	1.84 ^{c}	3.59 ^{a}	2.88 ^{b}	2.14 ^{bc}	2.12 ^{bc}	0.15	<0.001

Note: In the same row, values with different lower-case letters (a, b, c and d) indicate statistically significant differences ($p < 0.05$).

Regarding cytokines, IL-4 levels were higher in the AB, FC3, and FC1 groups, significantly surpassing ($p < 0.05$) those observed in the NC and PC groups. Conversely, TNF-α and IFN-γ levels were significantly reduced in the FC3 and AB groups (~41% and 20% lower than the PC group; $p < 0.05$).

The findings presented in Table 7 suggest that supplementation with fermented chive (FC1 and FC3) significantly influenced the gut microbial density in the *E. coli*-challenged chickens. Compared to the positive control (PC) group, the FC1, FC3, and antibiotic (AB) groups exhibited a significant reduction in *Salmonella* spp. counts ($p < 0.05$), ranging from 4.2 to 4.6 $\log_{10}$ CFU/g, compared to 6.7 $\log_{10}$ CFU/g in the PC group. Similarly, *E. coli* counts were also significantly reduced ($p < 0.05$) to 5.5–6.2 $\log_{10}$ CFU/g, compared to 7.1 $\log_{10}$ CFU/g in the PC group.

Table 7. Effects of fermented chive supplements on ileal microbial density in broiler chickens.

Bacteria	NC	PC	FC1	FC3	AB	Pooled SEM	*p*-Value
Salmonella spp.	4.61 ^{b}	6.68 ^{a}	4.63 ^{b}	4.21 ^{b}	4.54 ^{b}	0.34	<0.001
E. coli	6.67 ^{ab}	7.1 ^{a}	6.24 ^{b}	5.51 ^{c}	5.63 ^{bc}	0.48	<0.001
Lactobacillus spp.	4.84 ^{bc}	5.49 ^{b}	3.84 ^{c}	7.22 ^{a}	4.72 ^{bc}	0.79	<0.001

Note: In the same row, values with different lower-case letters (a, b and c) indicate statistically significant differences ($p < 0.05$).

Notably, the FC3 group demonstrated a remarkable increase in beneficial *Lactobacillus* spp. counts ($p < 0.05$), reaching 7.2 $\log_{10}$ CFU/g, which was significantly higher ($p < 0.05$) than the other groups (3.8–4.7 $\log_{10}$ CFU/g).

The histological analysis of the intestine (Table 8 and illustration Figure 4) demonstrated that across all intestinal segments (duodenum, jejunum, and ileum), FC supplementation, particularly at a 3% concentration (FC3), improved intestinal morphology by increasing VH and reducing CD across different intestinal segments. These effects were often comparable to the antibiotic group (AB) and, in some cases, approached the levels observed in the NC group.

Table 8. Effects of fermented chive supplementation on the intestinal histology (duodenum, jejunum, and ileum) of broiler chickens.

Items	NC	PC	FC1	FC3	AB	Pooled SEM	*p*-Value
Duodenum							
VH (μm)	2489.3 ^{a}	1817.5 ^{b}	2150.5 ^{ab}	2190.6 ^{ab}	2244.6 ^{ab}	110.8	0.005
CD (μm)	186.7 ^{c}	379.6 ^{a}	306.4 ^{b}	276.4 ^{b}	291.4 ^{b}	21.09	<0.001
VH:CD	13.33 ^{a}	4.79 ^{c}	7.02 ^{b}	7.93 ^{b}	7.70 ^{b}	0.71	<0.001
Jejunum							
VH (μm)	1304.1	1150.2	1182.6	1223	1277	46.8	0.119
CD (μm)	164.8 ^{b}	284.9 ^{b}	271.8 ^{a}	241.8 ^{a}	256.8 ^{a}	11.79	<0.001
VH:CD	7.91 ^{a}	4.04 ^{c}	4.39b ^{c}	5.06 ^{b}	4.97^{b}	0.39	<0.001
Ileum							
VH (μm)	902.7 ^{b}	786.9 ^{c}	936.8 ^{b}	1061.5 ^{a}	907.2 ^{b}	20.94	0.013
CD (μm)	152.8 ^{c}	206.44 ^{a}	184.69 ^{ab}	182.54 ^{ab}	171.65 ^{bc}	7.48	0.023
VH:CD	5.91 ^{a}	3.81 ^{c}	5.07 ^{ab}	5.82 ^{a}	5.29 ^{ab}	0.31	0.019

Note: In the same row, values with different lower-case letters (a, b and c) indicate statistically significant differences ($p < 0.05$); VH: villus height, CD: crypt depth.

Villus height (VH) was significantly reduced ($p < 0.001$) in the *E. coli*-challenged group (PC) compared to the negative control (NC) group. However, supplementation with fermented chive (FC1, FC3) and antibiotic treatment (AB) contributed to partial recovery of VH, particularly in the duodenum and ileum ($p < 0.05$). Crypt depth (CD) was notably increased in the PC group in all intestinal sections ($p < 0.001$), indicating potential villus damage and increased cellular turnover. FC supplementation, especially FC3, effectively alleviated this increase, leading to crypt depth values closer to those observed in the NC group. The villus height-to-crypt depth (VH:CD) ratio followed a similar pattern, with a significant reduction in the PC group across all intestinal segments ($p < 0.05$). This imbalance suggests compromised intestinal integrity and absorption capacity. Notably, FC supplementation, particularly FC3, contributed to restoring the VH:CD ratio, indicating improved intestinal health.

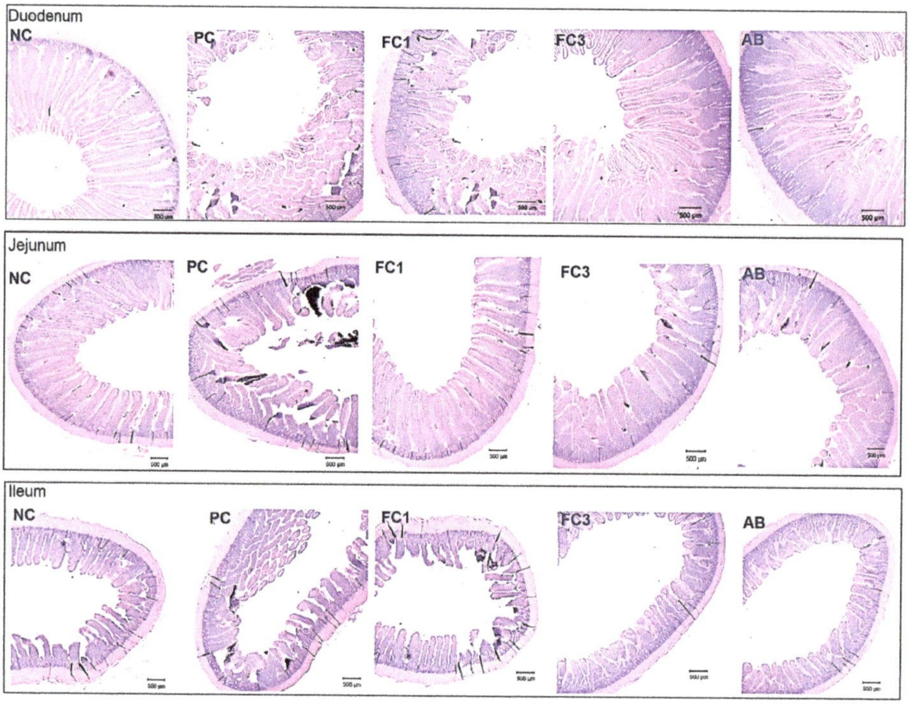

Figure 4. Effect of supplementation of polyherbal feed additive on the visualization of the histology of different intestinal segments (duodenum, jejunum, and ileum) by hematoxylin and eosin (H&E) in broiler chickens-NC (non-challenge control), basal diet with non-*E. coli* challenge; PC (positive control), basal diet with *E. coli* challenge; FC1, FC3, and AB-addition of 1% and 3% FC and Terra-Neocine to the basal diet with *E. coli* challenge.

4. Discussion

4.1. Effects of Supplementation on the Productivity of Experimental Broiler Chickens

After colonizing the intestines of young chickens, *E. coli* can lead to septicemia and systemic infections, negatively affecting gut health and growth performance [15]. The present study demonstrates that supplementation with FC, particularly 3% concentration (FC3), positively influences the growth performance of *E. coli*-challenged chickens. Specifically, the FC3 group exhibited a significant increase in body weight gain (BWG), feed intake (FI), and survival rate, reaching levels comparable to the antibiotic-treated group (AB).

Additionally, both the FC3 and FC1 groups showed an improved production efficiency index (PEI) compared to the positive control (PC) group. However, no significant differences in the feed conversion ratio (FCR) were observed among the groups. This could be attributed to the lower feed intake in the PC group due to disease effects, which resulted in FCR values not reflecting clear differences between treatments.

The observed performance improvements may be linked to bioactive compounds present in chive, such as polyphenols, flavonoids, and organosulfur compounds, known for their antioxidant, anti-inflammatory, and antimicrobial properties (Table 1). These compounds potentially inhibit *E. coli* proliferation in the gastrointestinal tract, thereby enhancing gut health and supporting growth [16]. The fermentation process appears to enhance the efficacy of these bioactive compounds while reducing anti-nutritional effects, optimizing health and growth benefits for chickens [17]. Interestingly, fermentation not only preserves active compounds but also generates by-products, such as short-chain fatty acids, which nourish epithelial cells and maintain intestinal mucosal integrity.

Furthermore, FC3 functions as a synbiotic, combining prebiotics (from chive) and probiotics (from *L. plantarum* fermentation). This synbiotic effect promotes the survival of beneficial gut bacteria, increases lactic acid production to lower intestinal pH, and inhibits pathogenic *E. coli* [18]. Chive also provides nutrients for beneficial bacteria, supporting microbiota balance, improving digestion, and enhancing growth performance [19].

Our results align with recent studies, such as those by Adli et al. (2024) [20], who reported improved broiler growth performance with fermented herbal supplements, and Liu et al. (2025) [21], who observed increased BWG and an improved FCR in early-stage broilers supplemented with lactic acid bacteria. Although the FCR did not show significant differences, the higher PEI in both the FC3 and FC1 groups compared to the PC group suggests that chive-based supplementation optimizes production efficiency. These findings highlight the potential of chive-derived supplements, particularly FC3, as antibiotic alternatives, which is especially relevant in the context of reducing antibiotic use to mitigate antimicrobial resistance.

4.2. Effects of Fermented Chive Supplementation on Immune Function

The thymus and bursa of Fabricius are central organs for T and B cell production, while the spleen serves as a peripheral organ containing abundant lymphocytes, supporting immune function in broilers [22]. The weight of immune organs reflects the immune health of chickens [23]. Increased bursal and splenic weights in the FC3 group may be attributed to the probiotic *L. plantarum* in the FC supplement, which stimulates lymphoid tissue development and supports immune function [24]. Polysaccharides from FC, combined with probiotics, may act as prebiotics, balancing gut microbiota and enhancing immune responses, similar to findings of Sugiharto and Ranjitkar (2019) [25].

Antibodies, or immunoglobulins, are essential glycoproteins produced by B lymphocytes, playing a crucial role in immune defense against bacterial and viral infections. This study evaluated the impact of *E. coli* challenge on serum immunoglobulin (Ig) levels in broilers. The results indicated a significant increase in IgA and IgM levels following supplementation. Ismail et al. (2021) [26] reported that garlic powder supplementation (0.75 g/kg feed) significantly increased IgM and IgG, enhancing both mucosal and systemic immunity; however, IgM—an early immune response antibody—showed no notable change with garlic supplementation. The PC group exhibited the lowest IgA and IgG levels, reflecting immunosuppression and a higher risk of intestinal infections.

Supplementation significantly modulated the expression of pro-inflammatory and anti-inflammatory cytokines, enhancing intestinal immune responses. Tight junction integrity and gut permeability are maintained by proteins such as occludin, ZO-1, and claudin. Wu et al. (2021) [27] demonstrated that *E. coli* O157 severely damages intestinal epithelial permeability and mucosal barrier function in broilers, allowing gut pathogens to translocate to the liver and bloodstream. Increased expression of ZO-1 and Claudin-2 in supplemented groups indicates improved epithelial integrity, consistent with Zhu et al. (2023) [28].

When a pathogen invades the colon and damages tissues, the intestinal epithelium releases inflammatory mediators, closely linked to gut inflammatory responses [29]. Elevated IL-4 levels in both the high-dose FC (FC3) and low-dose FC (FC1) groups indicate the activation of Th2 immune responses, aligning with Xu et al. (2023) [17]. Conversely, decreased IFN-γ and TNF-α levels in the supplemented groups suggest anti-inflammatory effects, whereas the PC group exhibited the highest IFN-γ and TNF-α levels, indicative of severe inflammation, particularly during early life stages [30].

Fermented chive increased IL-4 levels (by 83.68% and 69.52%) while reducing TNF-α and IFN-γ (~42% and 21%), demonstrating immunoregulatory effects comparable to antibiotics. Chive-derived supplements also promoted immune organ development and reduced mortality rates, enhancing humoral immune responses. These effects can be attributed to bioactive compounds, such as saponins, flavonoids, and polysaccharides. Additionally, *Lactobacillus* stimulates IL-4 production, enhances B cell development, and suppresses inflammation [31]. These findings suggest that fermented chive-based supplements have the potential to enhance immune health and protect chickens from *E. coli* infections.

4.3. Effects of Fermented Chive Supplementation on Gut Health

In this study, *E. coli*-infected broilers from the positive control (PC) group exhibited the highest mortality rates, indicating that *E. coli* challenge triggered inflammation and compromised gut health. The gut microbiota forms a multilayered bacterial barrier in the intestines, playing a crucial role in nutrient absorption, gut protection, and immune regulation [32]. The findings suggest that chive supplementation effectively controlled pathogenic bacteria, such as *E. coli* and *Salmonella* spp., while increasing beneficial *Lactobacillus* populations, particularly in the FC3 group. These results align with previous research demonstrating that herbal additives can inhibit harmful bacteria, selectively promote beneficial microbes, and regulate gut microbiota composition [33].

The gastrointestinal tract is a complex system in which maintaining structural and functional stability plays an important role in nutrient absorption, resistance to infection, and overall physiological performance [34]. The present study demonstrated that dietary FC supplementation, particularly at a 3% inclusion level (FC3), improved intestinal histology by increasing VH and reducing CD, leading to a higher VH:CD ratio compared to the positive control (PC) group with *E. coli*-induced intestinal damage.

The improvements in gut health observed in this study are likely due to the fermented chive's ability to inhibit *E. coli,* reduce intestinal damage, and promote recovery. The bioactive compounds in fermented chive, including antimicrobial and antioxidant agents, may enhance intestinal health by supporting epithelial cell turnover and maintaining the intestinal barrier. Lactic acid fermentation further contributes by producing active metabolites, such as propanoic acid, quercetin, and saponins, which possess antimicrobial and anti-inflammatory properties [16]. Additionally, *Lactobacillus* produces organic acids and other metabolites that lower intestinal pH, inhibit pathogenic bacteria, and support gut health [35].

The results of this study showed that the fermented chive improved intestinal morphology, thereby enhancing nutrient absorption efficiency. Chive-derived supplementation exhibited protective effects on intestinal mucosal morphology, significantly increasing VH while reducing CD, supporting improved gut integrity and nutrient absorption [36]. A higher VH:CD ratio may reduce the need to maintain intestinal structure, allowing more energy to be allocated to weight gain [37]. These mechanisms may explain how the fermented chive reduced the negative effects of *E. coli* on growth performance in the broiler chickens. Furthermore, maintaining intestinal structural integrity is closely linked to the stability of the intestinal immune system [38].

5. Conclusions

This study confirms the effectiveness of supplementing fermented chive with *L. plantarum* 1582 in improving performance, enhancing immunity, and protecting the gut microbiota of chickens infected with *E. coli* ExPEC_A338. The FC3 group achieved weight gain and feed intake levels comparable to those of the antibiotic (AB) and negative control (NC) groups, with significantly higher values than the positive control (PC) group. The mechanism of action of the supplement may involve immune enhancement (increased IgA and IgM levels, development of lymphoid organs), protection of the intestinal epithelial barrier (upregulation of ZO-1 and Claudin-2), and modulation of immune responses (increased IL-4, decreased IFN-γ and TNF-α). Additionally, the supplement inhibited pathogenic bacteria (*Salmonella* spp., *E. coli*), promoted beneficial bacteria (*Lactobacillus* spp.), and improved intestinal mucosal morphology. Fermented chive represents a promising antibiotic alternative in poultry farming, contributing to intestinal disease control and sustainable livestock production.

Author Contributions: Conceptualization, P.V.H.; methodology, P.V.H. and N.X.H.; formal analysis, L.X.A.; data curation, L.X.A.; writing—original draft preparation, P.V.H.; writing—review and editing, P.V.H. and N.X.H.; project administration, P.V.H. All authors have read and agreed to the published version of the manuscript.

Funding: This study was funded by the Vietnam Ministry of Education and Training under Project Code B2023-DHH-24.

Institutional Review Board Statement: All procedures related to the care, housing, and slaughter of experimental chickens were conducted in accordance with the standards approved by the Animal Ethics Advisory Committee, Hue University, Vietnam (Approval No.: HUVNO39 20 March 2024).

Informed Consent Statement: Not applicable.

Data Availability Statement: The raw data supporting the conclusions of this article will be made available by the authors on request.

Acknowledgments: We thank all of the members of the laboratory for their support and constructive comments. All authors included in this section have consented to the acknowledgement.

Conflicts of Interest: The authors declare no conflicts of interest.

References

1. Watts, A.; Wigley, P. Avian pathogenic *Escherichia coli*: An overview of infection biology, antimicrobial resistance and vaccination. *Antibiotics* **2024**, *13*, 809. [CrossRef]
2. Ronco, T.; Stegger, M.; Olsen, R.H.; Sekse, C.; Nordstoga, A.B.; Pohjanvirta, T.; Lilje, B.; Lyhs, U.; Andersen, P.S.; Pedersen, K. Spread of avian pathogenic *Escherichia coli* ST117 O78: H4 in Nordic broiler production. *BMC Genom.* **2017**, *18*, 1–8. [CrossRef]
3. Hai, P.V.; Hoa, N.X. Effect of Allium schoenoprasum extract on immune status against Newcastle virus and growth performance of broiler chicken. *Electron. J. Agric. Sci. Technol.—Hue Univ. Agric. For.* **2020**, *4*, 2058–2064.
4. Charen, E.; Harbord, N. Toxicity of herbs, vitamins, and supplements. *Adv. Chronic Kidney Dis.* **2020**, *27*, 67–71. [CrossRef] [PubMed]
5. Kim, S.; Park, S.-L.; Lee, S.; Lee, S.-Y.; Ko, S.; Yoo, M. UPLC/ESI-MS/MS analysis of compositional changes for organosulfur compounds in garlic (*Allium sativum* L.) during fermentation. *Food Chem.* **2016**, *211*, 555–559. [CrossRef] [PubMed]
6. Zhang, X.; Miao, Q.; Pan, C.; Yin, J.; Wang, L.; Qu, L.; Yin, Y.; Wei, Y. Research advances in probiotic fermentation of Chinese herbal medicines. *iMeta* **2023**, *2*, e93. [CrossRef] [PubMed]
7. Xiang, H.; Sun-Waterhouse, D.; Waterhouse, G.I.N.; Cui, C.; Ruan, Z. Fermentation-enabled wellness foods: A fresh perspective. *Food Sci. Hum. Wellness* **2019**, *8*, 203–243. [CrossRef]
8. Li, H.; Kang, Y.; Sun, Y.; Bian, C.; Fan, M.; Zhang, H.; Zhao, S.; Song, X.; Qiao, H. The role of Lactobacillus plantarum in solid-state fermentation of Astragalus membranaceus for broiler chicken feed. *AMB Express* **2025**, *15*, 26. [CrossRef]

9. Hai, P.; Phuong, H.F.C.; Hung, P.H.S.; Na, T.T.; Lai, N.H.; Khuong, N.D.T.; Liem, T.N.; Hoa, N.X. Selection of Lactobacillus strains from native chicken feces for the fermentation of purple onion (*Allium cepa* L.) as an antibiotic alternative against Salmonella spp. in chickens. *Open Vet. J.* **2024**, *14*, 3525–3538. [CrossRef]
10. *TCVN 11892-1:2017*; Good Agricultural Practice—Part 1: Crop Production. VietGAP: Hanoi, Vietnam, 2017.
11. *Complete Feed for Broilers—10 TCN 661-2005*; Vietnam Ministry of Agriculture and Rural Development: Hanoi, Vietnam, 2025.
12. Martins, J.M.S.; Carvalho, C.M.C.; Litz, F.; Silveira, M.M.; Moraes, C.; Silva, M.C.A.; Fagundes, N.; Fernandes, E. Productive and economic performance of broiler chickens subjected to different nutritional plans. *Rev. Bras. Cienc. Avic.* **2016**, *18*, 209–216. [CrossRef]
13. Livak, K.J.; Schmittgen, T.D. Analysis of relative gene expression data using real-time quantitative PCR and the 2(-Delta Delta C(T)) method. *Methods* **2001**, *25*, 402–408. [CrossRef]
14. Layton, C.; Bancroft, J.D.; Suvarna, S.K. 4—Fixation of tissues. In *Bancroft's Theory and Practice of Histological Techniques*, 8th ed.; Suvarna, S.K., Layton, C., Bancroft, J.D., Eds.; Elsevier: Amsterdam, The Netherlands, 2019; pp. 40–63.
15. Kathayat, D.; Lokesh, D.; Ranjit, S.; Rajashekara, G. Avian pathogenic Escherichia coli (APEC): An overview of virulence and pathogenesis factors, zoonotic potential, and control strategies. *Pathogens* **2021**, *10*, 467. [CrossRef] [PubMed]
16. Kothari, D.; Lee, W.-D.; Jung, E.; Niu, K.; Lee, C.; Kim, S.-K. Controlled fermentation using autochthonous Lactobacillus plantarum improves antimicrobial potential of Chinese chives against poultry pathogens. *Antibiotics* **2020**, *9*, 386. [CrossRef]
17. Xu, F.; Wu, H.; Xie, J.; Zeng, T.; Hao, L.; Xu, W.; Lu, L. The effects of fermented feed on the growth performance, antioxidant activity, immune function, intestinal digestive enzyme activity, morphology, and microflora of yellow-feather chickens. *Animals* **2023**, *13*, 3545. [CrossRef] [PubMed]
18. Alagawany, M.; Hegazy, M.I.; Farag, M.R.; El-Shall, N.A.; Sallam, S.M.; Dhama, K. Chapter 2—Probiotics, prebiotics, and synbiotics in animal and poultry nutrition. In *Organic Feed Additives for Livestock*; Elsevier: Amsterdam, The Netherlands, 2025; pp. 17–41.
19. Richards, P.J.; Almutrafy, A.; Liang, L.; Flaujac Lafontaine, G.M.; King, E.; Fish, N.M.; Connerton, A.J.; Connerton, P.L.; Connerton, I.F. Prebiotic galactooligosaccharide feed modifies the chicken gut microbiota to efficiently clear Salmonella. *mSystems* **2024**, *9*, e0075424. [CrossRef]
20. Adli, D.N.; Sholikin, M.M.; Ujilestari, T.; Ahmed, B.; Sadiqqua, A.; Harahap, M.A.; Sofyan, A.; Sugiharto, S. Effect of fermentation of herbal products on growth performance, breast meat quality, and intestinal morphology of broiler chickens: A meta-analysis. *Ital. J. Anim. Sci.* **2024**, *23*, 734–750. [CrossRef]
21. Liu, J.; Gu, H.; Jia, R.; Li, S.; Chen, Z.; Zheng, A.; Chang, W.; Liu, G. Effects of *Lactobacillus acidophilus* on production performance and immunity of broiler chickens and their mechanism. *Front. Vet. Sci.* **2025**, *12*, 1554502. [CrossRef]
22. Muratbayev, D.; Ygiyeva, A.; Bilyalov, Y.; Zaikovskaya, O.; Zhexenayeva, A. Morphogenesis of the spleen and cloacal bursa of a chicken embryo under the influence of "ligfolum" and "placenta denatured emulsified". *Int. J. Vet. Sci.* **2023**, *12*, 847–852.
23. Sattarova, R.; Shynybaev, K.; Bakiyeva, F.; Strochkov, V.; Boranbayeva, K.; Zhanserkenova, O.; Kassymbekova, S.; Ibadullayeva, A.; Khamzina, A. Metagenomic Analysis and identification of epizootic strains of the causative agent of infectious bovine Keratoconjunctivitis in Kazakhstan. *Int. J. Vet. Sci.* **2023**, *12*, 822–831.
24. Malematja, E.; Manyelo, G.; Sebola, A.; Mabelebele, M. The role of insects in promoting the health and gut status of poultry. *Comp. Clin. Pathol.* **2023**, *32*, 501–513. [CrossRef]
25. Sugiharto, S.; Ranjitkar, S. Recent advances in fermented feeds towards improved broiler chicken performance, gastrointestinal tract microecology and immune responses: A review. *Anim. Nutr.* **2019**, *5*, 1–10. [CrossRef] [PubMed]
26. Ismail, I.E.; Alagawany, M.; Taha, A.E.; Puvača, N.; Laudadio, V.; Tufarelli, V. Effect of dietary supplementation of garlic powder and phenyl acetic acid on productive performance, blood hematology, immunity and antioxidant status of broiler chickens. *Anim. Biosci.* **2021**, *34*, 363–370. [CrossRef] [PubMed]
27. Wu, Z.; Yang, K.; Zhang, A.; Chang, W.; Zheng, A.; Chen, Z.; Cai, H.; Liu, G. Effects of *Lactobacillus acidophilus* on the growth performance, immune response, and intestinal barrier function of broiler chickens challenged with *Escherichia coli* O157. *Poult. Sci.* **2021**, *100*, 101323. [CrossRef]
28. Zhu, X.; Tao, L.; Liu, H.; Yang, G. Effects of fermented feed on growth performance, immune organ indices, serum biochemical parameters, cecal odorous compound production, and the microbiota community in broilers. *Poult. Sci.* **2023**, *102*, 102629. [CrossRef]
29. Shang, Y.; Regassa, A.; Kim, J.H.; Kim, W.K. The effect of dietary fructooligosaccharide supplementation on growth performance, intestinal morphology, and immune responses in broiler chickens challenged with *Salmonella enteritidis* lipopolysaccharides. *Poult. Sci.* **2015**, *94*, 2887–2897. [CrossRef] [PubMed]

30. Paraschivescu, C.; Barbosa, S.; Van Steenwinckel, J.; Gressens, P.; Glaichenhaus, N.; Davidovic, L. Early Life Exposure to Tumor Necrosis Factor Induces Precocious Sensorimotor Reflexes Acquisition and Increases Locomotor Activity During Mouse Postnatal Development. *Front. Behav. Neurosci.* **2022**, *16*, 845458. [CrossRef] [PubMed]
31. Itoh, T.; Miyazono, D.; Sugata, H.; Mori, C.; Takahata, M. Anti-inflammatory Effects of heat-killed *Lactiplantibacillus argentoratensis* BBLB001 on a gut inflammation co-culture cell model and dextran sulfate sodium-induced colitis mouse model. *Int. Immunopharmacol.* **2024**, *143*, 113408. [CrossRef]
32. De Vos, W.M.; Tilg, H.; Van Hul, M.; Cani, P.D. Gut microbiome and health: Mechanistic insights. *Gut* **2022**, *71*, 1020–1032. [CrossRef]
33. Lin, T.-L.; Lu, C.-C.; Lai, W.-F.; Wu, T.-S.; Lu, J.-J.; Chen, Y.-M.; Tzeng, C.-M.; Liu, H.-T.; Wei, H.; Lai, H.-C. Role of gut microbiota in identification of novel TCM-derived active metabolites. *Protein Cell* **2021**, *12*, 394–410. [CrossRef]
34. Celi, P.; Cowieson, A.; Fru-Nji, F.; Steinert, R.; Kluenter, A.-M.; Verlhac, V. Gastrointestinal functionality in animal nutrition and health: New opportunities for sustainable animal production. *Anim. Feed. Sci. Technol.* **2017**, *234*, 88–100. [CrossRef]
35. Huang, R.; Wu, F.; Zhou, Q.; Wei, W.; Yue, J.; Xiao, B.; Luo, Z. Lactobacillus and intestinal diseases: Mechanisms of action and clinical applications. *Microbiol. Res.* **2022**, *260*, 127019. [CrossRef] [PubMed]
36. Guan, X.; Li, W.; Meng, H. A double-edged sword: Role of butyrate in the oral cavity and the gut. *Mol. Oral Microbiol.* **2021**, *36*, 121–131. [CrossRef] [PubMed]
37. Teng, P.-Y.; Yadav, S.; de Souza Castro, F.L.; Tompkins, Y.H.; Fuller, A.L.; Kim, W.K. Graded Eimeria challenge linearly regulated growth performance, dynamic change of gastrointestinal permeability, apparent ileal digestibility, intestinal morphology, and tight junctions of broiler chickens. *Poult. Sci.* **2020**, *99*, 4203–4216. [CrossRef] [PubMed]
38. Ceylan, A.; Saçaklı, P.; Özgenç Çınar, Ö.; Ramay, M.S.; Ahsan, U.; Harijaona, J.A.; Bayraktaroğlu, A.G.; Manghebati, F.; Calik, A. Effect of supplemental dietary phytogenic blends on growth performance, jejunal histomorphometry, and jejunal immunity of broiler chickens. *Arch. Anim. Breed.* **2025**, *68*, 13–26. [CrossRef]

fermentation

Article

Growth Performance and Rumen Microbiota of Sheep Respon… to Cotton Straw Fermented with Compound Probiotics

Peiling Wei [1], Mingxuan Guan [1], Xuhui Liang [2], Kaixin Yuan [2], Ning Chen [3], Yuxin Yang [2] and Ping Gong [1,*]

[1] Xinjiang Academy of Animal Science, Urumqi 830057, China; weipeiling0830@126.com (P.W.); 13639963750@163.com (M.G.)
[2] College of Animal Science and Technology, Northwest A&F University, Yangling 712100, China; liangxuhui@nwafu.edu.cn (X.L.); m15129066930@163.com (K.Y.); yangyuxin2002@126.com (Y.Y.)
[3] Institute of Animal Husbandry and Veterinary Medicine, Xinjiang Academy of Agricultural Reclamation Sciences, Shihezi 832000, China; ningningkuaile@126.com
* Correspondence: gongping2025@126.com

Abstract: To develop cotton straw as a feed resource through biological fermentation, it was fermented using compound probiotics (*Bacillus subtilis*, *Saccharomyces cerevisiae*, and *Lactobacillus plantarum*) and subsequently fed to sheep after the nutrients and hygienic indices of the fermented cotton straw (FCS) were analyzed. Sixty sheep were randomly assigned to five groups: a control group (CON); a low-proportion fermented cotton straw group (LFC, with FCS comprising 14.5% of the diet); a high-proportion fermented cotton straw group (HFC, with FCS comprising 29.0% of the diet); a compound microbial group (MIC, containing *Bacillus licheniformis*, *Bacillus subtilis*, and yeast); and a microbial-enzymatic preparation group (MEY, containing compound probiotics and enzymes such as cellulase, xylanase, β-glucanase, amylase, and protease). The trial lasted seven weeks and was divided into two stages: stage 1 (weeks 1–4, days 1–28) and stage 2 (weeks 5–7, days 29–49). Body weight and daily feed intake were registered, and blood and rumen fluid samples were obtained at day 28 and day 49 of the feeding trial. Fermentation significantly increased the crude protein content of cotton straw while reducing neutral detergent fiber (NDF) and acid detergent fiber (ADF) ($p < 0.05$). Additionally, fermentation reduced the residues of aflatoxin B1, vomitoxin, zearalenone, and free gossypol in the treatment groups ($p < 0.05$). LFC possessed the lowest value of feed-to-gain ratio (F/G) among all groups. Serum indices related to antioxidant capacity and utilization of fat and protein increased in the treatment group ($p < 0.05$). Rumen microbiota were separated between different groups ($p < 0.05$). LFC and HFC enhanced the abundance of Prevotella. These findings could provide conclusions that fermented cotton straw has the tendency to enhance the growth performance of sheep by increasing the abundance of bacteria related to utilization of protein, carbohydrate, and other nutrients such as *Prevotella*, in which the LFC group has the best fast-fattening (about 50 d) effect.

Keywords: biological fermentation; cotton straw; sheep; probiotics

1. Introduction

Cotton is a crucial source of natural fiber for the global textile industry [1]. Cotton cultivation spans over 30 million hectares across 85 countries and regions, with more than half participating in the global cotton textile trade [2]. Cotton straw is the main by-product of cotton; however, vast amounts of cotton straw are burned, which not only causes the waste of resources but also aggravates environmental pollution because of the SO_2 and NO_2 produced [3]. Cotton straw is rich in nutrients, and its crude protein (6.61%), calcium (0.34%), and phosphorus (0.15%) contents are higher than those of corn straw, rice straw, and wheat straw; hence, it possesses the potential to be a high-quality feed resource. Efficient use of cotton straw in animal husbandry could reduce carbon emissions and provide conditions for the production of green agricultural products, which is in line with the concept of green ecological development of animal husbandry [4,5]. However, cotton straw is hard in texture and poor in taste, so direct-feeding efficiency is unsatisfactory. Straw could be treated by physical methods such as steam explosion, chemical methods such as ammoniation, and biological methods such as fermentation then fed to cattle and sheep [6–8]. Among these methods, the bacterial-enzyme synergistic biological fermentation not only enhances the nutritional value of feed but also improves its palatability, addressing the limitations of physical and chemical methods. The application of probiotics as fermentative inoculants in feed production has been shown to markedly enhance both feed quality and the antioxidant capacity of the resulting product [9–12]. Research indicates that aerobic bacilli consume free oxygen in the intestinal tract during their growth and metabolic processes, creating an anaerobic environment that supports the growth, reproduction, and metabolism of anaerobic probiotics like lactic acid bacteria [13]. Lactic acid bacteria, such as *Lactobacillus plantarum*, exhibit strong antibacterial activity against a wide range of foodborne pathogens [14]. Yeasts, including Saccharomyces cerevisiae, inhibit mold growth in feed [15]. Additionally, cellulase and protease produced by probiotics enhance nutrient utilization, while enzymes that degrade mycotoxins and gossypol safeguard animal health [10,16,17]. Therefore, this study seeks to develop cotton straw as a novel feed resource through biological fermentation and to evaluate its feeding effects in sheep. The findings aim to offer theoretical and practical insights into the development and utilization of innovative feed resources for ruminants.

2. Materials and Methods

2.1. Production of Fermented Cotton Straw

The probiotics employed in the biological fermentation process, including *Bacillus subtilis*, *Lactobacillus plantarum*, and *Saccharomyces cerevisiae*, were preserved and supplied by our research team.

The cotton straw was provided by Tumushuke Wanzhiyang Farmers Professional Cooperative. The inoculation ratio of *Lactobacillus plantarum/Bacillus subtilis/Saccharomyces cerevisiae* = 2:1:1 was used to produce fermented cotton straw (inoculum dose: 1% of the total weight of cotton straw). The fermentation of straw was carried out using a mixed inoculum containing *Bacillus subtilis* ($\geq 1.0 \times 10^9$ CFU/g), *Lactobacillus plantarum* ($\geq 0.5 \times 10^9$ CFU/g), and *Saccharomyces cerevisiae* ($\geq 1.0 \times 10^9$ CFU/g). The microbial strains were preserved and provided by the Sheep Genetic Improvement and Germplasm Innovation Team of the College of Animal Science and Technology, Northwest A&F University. The cotton straw fermented by activated probiotics was the treatment group (T), the straw fermented by inactivated probiotics (autoclaved, 127 °C) was the control group (C), and the cotton straw without treatment was the raw material (R). Feed samples were randomly collected on days 0, 14, 28, 42, and 56. The first aliquot of samples was for sensory evaluation and measurement of feed hygienic indices, and the second aliquot of the samples was used for measurement of nutritional components. After evaluation of feed hygiene and the nutrient level of the fermented cotton straw (FCS), large-scale production of fermented cotton straw was made for a growth trial of sheep.

2.2. Experiment Animals and Design

The trial was conducted at the Tumushuke Wanzhiyang Farmers Professional Cooperative in Xinjiang, China (39.91 N 79.51 E). The feeding period was 11 weeks in total, including 2 weeks for transition, 2 weeks for adaptation, and 7 weeks for the growth trial. The growth trial was divided into 2 stages: stage 1 was week 1–week 4 (day 1–day 28), and stage 2 was week 5–week 7 (day 29–day 49). Sixty 4-month-old sheep (Duolang sheep♂ × Karakul sheep♀) in healthy condition (BW 28.82 ± 3.35 kg) were randomly divided into 5 groups (12 sheep/group). Each group consisted of four replicates, with three sheep per replicate housed in the same sheep shed, sharing one pen per replicate. The groups were CON (control group), LFC (low-proportion fermented cotton straw group, where FCS accounted for 14.50% of the diet), HFC (high-proportion fermented cotton straw group, where FCS accounted for 29.00% of the diet), MIC (microbial group) and MEY (mixture of microbial and enzymatic preparation group). The sheep were provided by Tumushuke Wanzhiyang Farmers Professional Cooperative. The probiotics used in the MIC group were developed with the participation of the Sheep Genetic Improvement and Germplasm Innovation Team of the College of Animal Science and Technology, Northwest A&F University, while the formulation used in the MEY group was provided by Guangdong Vtr Bio-Tech Co., Ltd. Before the start of the trial, all sheep pens were thoroughly disinfected, and the sheep were dewormed and numbered. Animals were fed twice a day at 8:00 and 18:00, and all the animals were provided with free access to feed and water.

2.3. Diets

All diets were formulated according to the feeding standards of meat-producing sheep and goats. There are 3 kinds of diet in this study shown in Table 1: a basal diet for the CON group, a low-proportion fermented cotton straw diet for the LFC group, and a high-proportion fermented cotton straw diet for the HFC group. On the basis of the CON group, the MIC group was given compound probiotics (2% of the daily feed); the MEY group was given an enzyme and *Saccharomyces cerevisiae* combination preparation (1% of the daily feed) on the basis of the MIC group. The MIC used in this study was developed by our research team and subsequently commercialized by First Bio-technology Co., Ltd. (Xianyang, China). The composition of MIC included *Bacillus licheniformis* ($\geq 1.0 \times 10^8$ CFU/g), *Bacillus subtilis* ($\geq 5.0 \times 10^7$ CFU/g), and *Saccharomyces cerevisiae* ($\geq 1.0 \times 10^8$ CFU/g). Regarding the MEY compound preparation, it contained Saccharomyces cerevisiae (dry matter) at $\geq 1.0 \times 10^9$ CFU/g, with enzyme activities as follows: cellulase ≥ 3000 U/g, xylanase ≥ 2000 U/g, β-glucanase ≥ 15,000 U/g, amylase ≥ 20,000 U/g, and protease ≥ 2000 U/g, along with various other fermentation-derived metabolites. This preparation was developed by VTR Bio-Tech Co., Ltd. (Zhuhai, China).

Table 1. The composition of diet (dry matter basis) %.

Items	Groups [2]		
	CON	LFC	HFC
Ingredients			
Cotton leaf	43.50	29.00	14.50
Fermented cotton straw	0	14.50	29.00
Silage corn	16.90	16.90	16.90
Corn	21.70	21.70	21.70
Soybean meal	9.35	9.35	9.35
Cottonseed meal	5.75	5.75	5.75
$CaHPO_4$	0.80	0.80	0.80
$NaHCO_3$	0.50	0.50	0.50

Table 1. *Cont.*

Items	Groups [2]		
	CON	LFC	HFC
NaCl	0.50	0.50	0.50
Premix [1]	1.00	1.00	1.00
Total	100.00	100.00	100.00
Nutrient levels [3]			
CP	14.62	13.65	12.66
CF	19.14	21.87	23.35
EE	4.36	3.61	2.65
Ca	0.81	0.93	0.98
P	0.46	0.49	0.50

[1] The premix was Cu 200 mg, Zn 586 mg, Mn 360 mg, Fe 765 mg, vitamin A 100,000 IU, vitamin D 50,000 IU, and vitamin E 380 IU. [2] CON: control group with basal diet; LFC: low-proportion fermented cotton straw diet (FCS accounted for 14.50% of the diet) group; HFC: high-proportion fermented cotton straw diet (FCS accounted for 29.00% of the diet) group. [3] CP, crude protein; CF, crude fiber; EE, ether extract; Ca, calcium; P, phosphorus.

2.4. Feed Proximate Analyses and Measurement of Hygienic Indices

Analyses performed on feed samples were as follows: (1) The crude protein content of the diet was tested using the Kjeldahl method [18]. (2) The Van Soest detergent fire analysis method was used for the determination of neutral detergent fiber (NDF) and acid detergent fiber (ADF) [19]. (3) The contents of free-gossypol (FG) were determined by the aniline method [20]. (4) Aflatoxin B1 (AFB1), vomitoxin (DON) and zearalenone (ZEN) were measured by ELISA kit (Xiamen lunchangshuo Biotechnology Co., Ltd., Xiamen, China). (5) The pH of the rumen fluid was immediately measured using a calibrated pH meter (Mettler Toledo Instruments Co., Ltd., Shanghai, China).

2.5. Feed Trial

The body weight of each sheep was assessed before morning feeding on day 1, day 28, and day 49 of the growth trial, and the feed intake was registered daily in order to calculate the ratio of feed to gain (F/G). Blood samples were collected from the jugular vein, 10 mL for each sheep, and centrifuged with 3000 r/min for 15 min to collect serum on day 28 and day 49 of the growth trial for the serum indices measurement. The contents of total cholesterol (TC), triacylglycerol (TG), urea (UREA), acetyl coenzyme A (AcCoA), beta-hydroxybutyric acid (BHBA), malondialdehyde (MDA), superoxide dismutase (SOD), and total antioxidant capacity (T-AOC) were measured by the kits (Sino-UK Institute of Biological Technology, Beijing, China).

2.6. Rumen Microbiota Diversity

Rumen fluid samples were collected by the oral intubation method from each sheep on day 28 and day 49 of the growth trial for the rumen microbial diversity measurement. The microbial DNA extraction was carried out using the E.Z.N.A.® soil DNA kit (Omega Bio-tek, Norcross, GA, USA). Nucleic acid quantifier NanoDrop2000 (Life Technologies, 109, Carlsbad, CA, USA) was used to determine the concentration and purity of the extracted DNA, while its quality was assessed using 1% agarose gel electrophoresis. The DNA samples that met the requirements were subjected to MiSeq sequencing. The bacterial 16S rRNA gene universal primers 338F (5′-ACTCCTACGGGAGGCAGCAG-3′) and 806R (5′-GGACTACHVGGGTWTCTAAT-3′) were used to amplify the original sequence. The PCR products were pooled with equal molar amounts from different samples. Sequencing libraries were generated using NEXTFLEX® Rapid DNA-Seq (Bioo Scientific, Austin, TX, USA) Kit. Fastp(https://github.com/OpenGene/fastp accessed on 26 August 2024 version 0.19.6) software was used to control the quality of the raw sequences, and FLASH (http://www.cbcb.umd.edu/software/flash accessed on 26 August 2024, version 1.2.11) software was used for splicing. The DADA2 plug-in in the QIIME2 process was used to denoise the optimized sequence after quality control. Sequences after DADA2 denoising were often referred to as ASVs (amplicon sequence variants). Follow-up data analysis based on ASVs was performed using the free online platform of the Majorbio Cloud Platform (www.majorbio.com) from Majorbio Bio-pharm Technology Co., Ltd. (Shanghai, China).

2.7. Statistical Analyses

All the data were preliminarily processed by EXCEL (Version 2021, Microsoft, Redmond, WA, USA). The data of feed nutrients, feed hygienic indices, growth performance, serum indices, and rumen fluid pH of sheep were analyzed using one-way analysis of variance by R software (Version 4.1.3, R Core Team, Vienna, Austria). The normality of the data was checked with the "qqPlot" function in "car" package, and the homogeneity of variance was checked by the "bartlett. test" function in the "stat" package. The "aov" function was subjected to perform ANOVA, and the "Duncan. test" function in the "agricolae" package was used to perform post hoc comparisons. Sequencing data were analyzed on the Majorbio platform (https://cloud.majorbio.com/). The test results were expressed in the form of "Mean ± SD"; $p < 0.05$ indicated significant differences.

3. Results

3.1. Effect of Fermented Cotton Straw with Compound Probiotics

Nutrient levels of cotton straw before and after fermentation are presented in Table 2. At the end of the fermentation, the contents of CP in both the treatment group and the control group were significantly higher than that of the raw material ($p < 0.05$), among which the treatment group exhibited a higher level. The levels of NDF and ADF in the treatment group and the control group were significantly lower than that of the raw material ($p < 0.05$), among which the treatment group exhibited a lower level.

Table 2. Changes of nutrient components of cotton straw with fermentation time (DM basis, $n = 6$), %.

Items [1]	Sampling Time					*p*-Value
	Day 0	Day 14	Day 28	Day 42	Day 56	
Control						
CP	6.74 ± 0.02 [b]	6.87 ± 0.14 [b]	6.92 ± 0.03 [b]	7.20 ± 0.14 [a]	7.18 ± 0.08 [a]	0.001
NDF	73.28 ± 2.14 [a]	67.58 ± 3.14 [b]	61.79 ± 1.70 [c]	64.65 ± 2.63 [bc]	64.15 ± 1.57 [bc]	<0.001
ADF	62.41 ± 1.69 [a]	56.49 ± 2.68 [ab]	51.04 ± 1.29 [b]	54.03 ± 1.93 [b]	53.27 ± 0.84 [b]	<0.001
Treatment						
CP	6.74 ± 0.02 [b]	7.92 ± 0.06 [a]	8.07 ± 0.05 [a]	8.30 ± 0.18 [a]	8.20 ± 0.01 [a]	<0.001
NDF	73.28 ± 2.14 [a]	65.96 ± 5.32 [b]	66.14 ± 2.01 [b]	61.05 ± 2.37 [c]	60.68 ± 1.23 [c]	<0.001
ADF	62.41 ± 1.69 [a]	57.19 ± 1.25 [b]	56.87 ± 1.11 [b]	52.43 ± 1.42 [c]	51.54 ± 0.30 [c]	<0.001

[1] CP, crude protein; NDF, neutral detergent fiber; ADF, acid detergent fiber. Note: In the same row, values with different letter superscripts mean significant differences ($p < 0.05$).

The residues of aflatoxin B1 (AFB1), vomitoxin (DON), zearalenone (ZEN), and free gossypol (FG) are shown in Figure 1. The amounts of AFB1 residue in the raw material, control group, and treatment group were 0.388 ppb, 0.260 ppb, and 0.281 ppb, respectively. Compared with the raw material, the treatment group decreased by 27.58% ($p < 0.05$) (Figure 1A); the amounts of DON residue in the raw material, control group, and treatment group were 245.203 ppb, 212.863 ppb, and 119.998 ppb, respectively. Compared with the raw material, the treatment group reduced by 51.06% ($p < 0.05$) (Figure 1B); the amounts of ZEN residue in the raw material, control group, and treatment group were 2.750 ppb, 2.218 ppb, and 2.145 ppb, respectively. Compared with the raw material, the treatment group dropped by 22.00% ($p < 0.05$) (Figure 1C); the amounts of FG residue in the raw material, control group, and treatment group were 203.825 mg/kg, 141.520 mg/kg, and 110.413 mg/kg, respectively. Compared with the raw material, the treatment group decreased by 45.83% ($p < 0.05$) (Figure 1D).

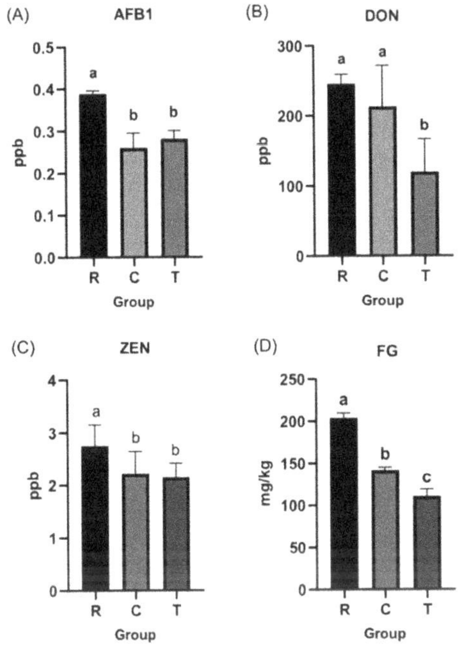

Figure 1. Hazardous substances residues in different groups. T, treatment group, cotton straw fermented by activated probiotics for 56 d; C, control group, cotton straw fermented by inactivated probiotics for 56 d; R, raw material, cotton straw without treatment. (**A**) Aflatoxin B1, AFB1 ($p < 0.001$); (**B**) vomitoxin, DON ($p < 0.001$); (**C**) zearalenone, ZEN ($p = 0.004$); (**D**) free gossypol, FG ($p < 0.001$). ($n = 6$). Note: In the same row, values with different letter superscripts mean significant differences ($p < 0.05$).

3.2. *Effects of Different Diets on Growth Performance of Sheep*

The growth performance of sheep under different diets is presented in Table 3. At the beginning of the trial, the average weight of sheep in each group was similar, with no significant differences observed ($p < 0.05$). During stage 1 and the entire trial period, the dry matter intake (DMI) in the HFC group was significantly lower than that of the other groups ($p < 0.05$). In stage 2, the DMI in the HFC group remained significantly lower than that of the CON, MIC, and MEY groups ($p < 0.05$). Additionally, in stage 2, the LFC group showed a tendency to increase the average daily gain (ADG) and to reduce the feed-to-gain ratio (F/G).

Table 3. Effect of different diets on growth performance of sheep ($n = 12$).

Items [1]	Groups [2]					*p*-Value
	CON	MIC	MEY	LFC	HFC	
BW (kg)						
Initial BW	28.77 ± 3.30	28.73 ± 2.94	29.03 ± 4.26	28.43 ± 4.06	29.15 ± 2.38	0.988
Final BW	38.41 ± 4.42	39.12 ± 3.29	39.55 ± 4.14	39.26 ± 4.02	38.45 ± 3.16	0.933

Table 3. *Cont.*

Items [1]	Groups [2]					p-Value
	CON	MIC	MEY	LFC	HFC	
DMI (g/d)						
W1–W4	1226.05 ± 60.07 ^{a}	1246.93 ± 33.49 ^{a}	1222.94 ± 33.29 ^{a}	1188.84 ± 41.00 ^{a}	1118.70 ± 43.88 ^{b}	0.007
W5–W7	1334.50 ± 108.05 ^{a}	1365.82 ± 75.24 ^{a}	1315.80 ± 77.55 ^{a}	1276.57 ± 77.47 ^{ab}	1155.61 ± 107.32 ^{b}	0.041
W1–W7	1273.50 ± 77.03 ^{a}	1298.94 ± 51.71 ^{a}	1263.56 ± 45.67 ^{a}	1227.22 ± 50.34 ^{a}	1134.85 ± 71.19 ^{b}	0.014
ADG (g/d)						
W1–W4	243.21 ± 72.08	265.59 ± 18.05	263.12 ± 46.57	239.35 ± 50.67	225.00 ± 28.71	0.720
W5–W7	146.43 ± 36.43	152.98 ± 26.26	162.70 ± 31.07	208.33 ± 32.16	153.57 ± 17.68	0.059
W1–W7	200.87 ± 34.07	216.32 ± 14.51	219.18 ± 31.32	225.78 ± 34.29	193.75 ± 22.09	0.494
F/G						
W1–W4	5.36 ± 1.45	4.71 ± 0.24	4.74 ± 0.68	5.16 ± 1.22	5.03 ± 0.61	0.803
W5–W7	9.55 ± 2.39	9.15 ± 1.81	8.30 ± 1.58	6.21 ± 0.75	7.54 ± 0.22	0.056
W1–W7	6.43 ± 0.74	6.02 ± 0.35	5.83 ± 0.65	5.53 ± 0.85	5.89 ± 0.42	0.681

[1] BW, body weight; DMI, dry matter intake; ADG, average daily gain; F/G, the ratio of feed to gain. [2] CON: control group with basal diet; MIC: added compound probiotics (2% of the feed amount) on the basis of CON; MEY: added enzyme and yeast combination preparation (1% of the feed amount) on the basis of MIC; LFC: low-proportion fermented cotton straw (FCS accounted for 14.50% of the diet) group; HFC: high-proportion fermented cotton straw (FCS accounted for 29.00% of the diet) group. Note: In the same row, values with different letter superscripts mean significant differences ($p < 0.05$).

3.3. *Effects of Different Diets on Serum Biochemical Parameters of Sheep*

The serum biochemical parameters of the sheep are presented in Table 4. In stage 1, the AcCoA levels were significantly increased in the MIC and LFC groups ($p < 0.05$). Growth hormone (GH) levels differed significantly among the groups ($p < 0.05$). The MIC, LFC, and HFC groups all exhibited higher GH levels compared to the control group, with the LFC group showing the highest value. Both MIC and LFC groups increased the SOD level ($p < 0.05$), and LFC enhanced the T-AOC level ($p < 0.05$). Among the groups, LFC had the highest levels of AcCoA, GH, SOD, and T-AOC. Furthermore, both MIC and LFC decreased the MDA level, with MIC showing the lowest value.

Table 4. Effect of different diets on serum biochemical parameters of sheep (n = 12).

Items [1]	Groups [2]					p-Value
	CON	MIC	MEY	LFC	HFC	
W1–W4 mmol/L						
TC	1.34 ± 0.33	1.13 ± 0.35	1.25 ± 0.50	1.00 ± 0.40	1.52 ± 0.66	0.085
TG	0.31 ± 0.14	0.28 ± 0.12	0.38 ± 0.16	0.28 ± 0.13	0.44 ± 0.35	0.224
UREA	5.10 ± 1.83	4.49 ± 1.44	5.07 ± 1.41	4.43 ± 1.26	6.01 ± 1.53	0.089
AcCoA	36.76 ± 6.99 ^{bc}	43.76 ± 2.11 ^{a}	33.75 ± 3.94 ^{c}	48.50 ± 8.32 ^{a}	40.13 ± 3.16 ^{b}	<0.001
BHBA	0.30 ± 0.18	0.30 ± 0.06	0.41 ± 0.15	0.29 ± 0.11	0.36 ± 0.17	0.210
GH (ng/mL)	4.91 ± 0.36 ^{d}	5.98 ± 0.38 ^{b}	4.46 ± 0.50 ^{e}	6.52 ± 0.49 ^{a}	5.51 ± 0.53 ^{c}	<0.001
SOD (U/mL)	64.50 ± 16.08 ^{bc}	73.94 ± 4.53 ^{a}	58.84 ± 3.18 ^{c}	75.85 ± 2.70 ^{a}	67.92 ± 2.10 ^{b}	<0.001
MDA (nmol/mL)	4.34 ± 0.63 ^{b}	3.13 ± 0.72 ^{d}	4.99 ± 0.38 ^{a}	3.50 ± 0.93 ^{cd}	4.05 ± 0.70 ^{bc}	<0.001
T-AOC (U/mL)	8.14 ± 1.19 ^{bc}	9.34 ± 0.62 ^{ab}	7.65 ± 1.10 ^{c}	10.99 ± 1.73 ^{a}	8.86 ± 0.42 ^{b}	<0.001
W5–W7 mmol/L						
TC	1.47 ± 0.41 ^{a}	0.60 ± 0.32 ^{bc}	0.41 ± 0.23 ^{c}	0.69 ± 0.32 ^{bc}	0.77 ± 0.32 ^{b}	<0.001
TG	0.30 ± 0.10 ^{a}	0.13 ± 0.09 ^{bc}	0.09 ± 0.06 ^{c}	0.17 ± 0.10 ^{b}	0.16 ± 0.08 ^{b}	<0.001
UREA	5.69 ± 1.25 ^{a}	3.57 ± 1.11 ^{cd}	3.03 ± 0.72 ^{d}	4.00 ± 1.04 ^{bc}	4.51 ± 1.01 ^{b}	<0.001
AcCoA	29.98 ± 7.25 ^{d}	56.21 ± 3.55 ^{b}	66.00 ± 4.21 ^{a}	50.66 ± 5.25 ^{c}	53.25 ± 5.66 ^{bc}	<0.001
BHBA	0.33 ± 0.11 ^{a}	0.18 ± 0.08 ^{b}	0.14 ± 0.09 ^{b}	0.21 ± 0.10 ^{b}	0.17 ± 0.06 ^{b}	<0.001
GH (ng/mL)	3.49 ± 0.58 ^{d}	8.10 ± 0.47 ^{b}	9.32 ± 0.39 ^{a}	7.59 ± 0.35 ^{c}	8.48 ± 0.54 ^{b}	<0.001
SOD (U/mL)	52.30 ± 2.66 ^{d}	84.06 ± 2.71 ^{b}	89.16 ± 2.81 ^{a}	78.10 ± 3.93 ^{c}	81.64 ± 3.11 ^{b}	<0.001
MDA (nmol/mL)	5.58 ± 0.91 ^{a}	2.11 ± 0.44 ^{c}	1.25 ± 0.40 ^{d}	2.85 ± 0.22 ^{b}	2.48 ± 0.30 ^{c}	<0.001
T-AOC (U/mL)	6.64 ± 0.46 ^{d}	13.74 ± 1.49 ^{ab}	14.65 ± 1.03 ^{a}	11.61 ± 1.04 ^{c}	12.84 ± 1.56 ^{bc}	<0.001

[1] TC, total cholesterol; TG, triglyceride; AcCoA, acetyl coenzyme A; BHBA, beta-hydroxybutyric acid; GH, growth

hormone; SOD, superoxide dismutase; MDA, malonaldehyde; T-AOC, total antioxidant. [2] CON: control group with basal diet; MIC: added compound probiotics (2% of the feed amount) on the basis of CON; MEY: added enzyme and yeast combination preparation (1% of the feed amount) on the basis of MIC; LFC: low-proportion fermented cotton straw (FCS accounted for 14.50% of the diet) group; HFC: high-proportion fermented cotton straw (FCS accounted for 29.00% of the diet) group. Note: In the same row, values with different letter superscripts mean significant differences ($p < 0.05$).

In stage 2, all treatment groups significantly reduced the levels of TC, TG, UREA, BHBA, and MDA ($p < 0.05$), with the MEY group showing the lowest values. Additionally, all treatment groups significantly increased the levels of AcCoA, GH, SOD, and T-AOC ($p < 0.05$), with the MEY group showing the highest levels.

3.4. Effects of Different Diets on Rumen Fluid pH of Sheep

The rumen fluid pH values for the different groups are presented in Table 5. In stage 1, the MIC and HFC groups significantly increased the pH values ($p < 0.05$), with the pH fluctuating between 6.15 and 6.66. In stage 2, no significant differences were observed between the groups, with pH values ranging from 6.60 to 6.86.

Table 5. Effect of different diets on rumen fluid pH of sheep ($n = 12$).

Items	Groups [1]					*p*-Value
	CON	MIC	MEY	LFC	HFC	
W1–W4	6.20 ± 0.33^{b}	6.60 ± 0.17^{a}	6.15 ± 0.26^{b}	6.18 ± 0.24^{b}	6.66 ± 0.16^{a}	<0.001
W5–W7	6.60 ± 0.16	6.71 ± 0.14	6.86 ± 0.28	6.77 ± 0.35	6.73 ± 0.45	0.111

[1] CON: control group with basal diet; MIC: added compound probiotics (2% of the feed amount) on the basis of CON; MEY: added enzyme and yeast combination preparation (1% of the feed amount) on the basis of MIC; LFC: low-proportion fermented cotton straw (FCS accounted for 14.50% of the diet) group; HFC: high-proportion fermented cotton straw (FCS accounted for 29.00% of the diet) group. Note: In the same row, values with different letter superscripts mean significant differences ($p < 0.05$).

3.5. Effects of Different Diets on Rumen Microbiota Diversity of Sheep

All 120 rumen fluid samples were subjected to 16S rRNA sequencing. An amount of 7,549,256 original sequences were obtained with 2,264,776,800 bp. After denoising by DADA2, a total of 6,679,844 optimized sequences were obtained with 2,794,570,466 bases (average sequence length of 418 bp). After smoothing according to the minimum number of sample sequences, 24,350 effective sequences for each sample were obtained. The statistics of the taxonomy annotation results were as follows: 24 phylum (phyla), 45 classes (class), 104 orders (order), 173 families (family), 374 genera (genus), 863 species (species), and 18,684 amplicon sequence variants (amplicon sequence variant, ASV). Subsequent microbial diversity data analyses were based on the ASVs dataset. The rarefaction curve showed that all the samples were sequenced with sufficient depth and reached the plateau (Figure 2).

The indices related to the alpha diversity of rumen microflora are listed in Table 6. In stage 1, no significant difference was observed. All treatment groups had the tendency to elevate the levels of Sobs, Ace, and Chao 1, and HFC had the most obvious impact. MIC and HFC tended to enhance the levels of Shannon and Shannoneven, among which HFC and MIC had the highest levels. In stage 2, MIC and MEY enhanced the level of coverage ($p < 0.05$). LFC and HFC enhanced the levels of Ace and Chao 1 ($p < 0.05$). All treatment groups tended to increase the levels of Shannon and Shannoneven.

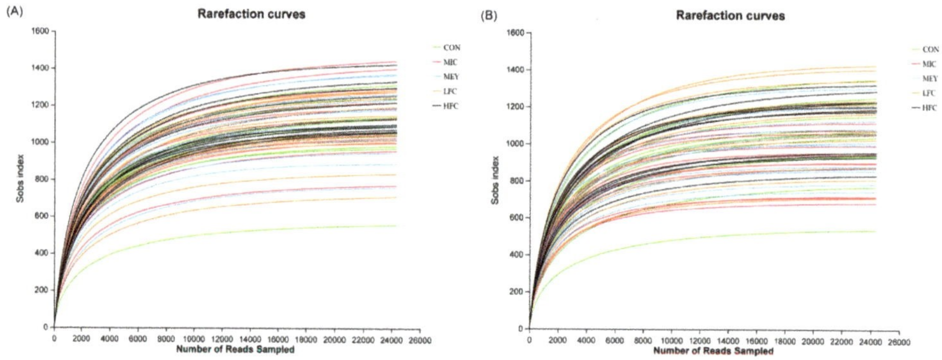

Figure 2. Rarefaction curve of rumen microflora. (**A**) Stage 1 (W1-W4); (**B**) stage 2 (W5-W7) (n = 12).

Table 6. Effects of different diets on alpha-diversity of rumen microflora of mutton sheep (n = 12).

Items	Groups [1]					p-Value
	CON	MIC	MEY	LFC	HFC	
W1–W4						
Coverage	0.9975 ± 0.008	0.9974 ± 0.0010	0.9977 ± 0.0010	0.9977 ± 0.0010	0.9973 ± 0.0006	0.731
Sobs	1069.00 ± 191.39	1146.42 ± 198.49	1113.83 ± 189.02	1089.50 ± 184.17	1177.75 ± 124.18	0.588
Ace	1090.05 ± 195.66	1168.28 ± 204.11	1132.97 ± 196.88	1108.59 ± 189.31	1200.35 ± 125.38	0.594
Chao1	1089.71 ± 195.93	1167.73 ± 207.02	1131.35 ± 197.69	1106.45 ± 189.51	1199.29 ± 124.66	0.596
Shannon	5.86 ± 0.49	6.00 ± 0.33	5.86 ± 0.48	5.87 ± 0.33	6.04 ± 0.36	0.693
Shannoneven	0.841 ± 0.048	0.864 ± 0.028	0.834 ± 0.049	0.841 ± 0.030	0.854 ± 0.042	0.735
W5–W7						
Coverage	0.9977 ± 0.0007 [c]	0.9996 ± 0.0003 [a]	0.9990 ± 0.0005 [b]	0.9981 ± 0.0006 [c]	0.9982 ± 0.0007 [c]	<0.001
Sobs	1033.00 ± 221.07	910.17 ± 145.13	992.00 ± 170.44	1121.33 ± 232.48	1103.92 ± 167.43	0.057
Ace	1052.12 ± 225.00 [ab]	912.39 ± 145.05 [b]	998.69 ± 172.23 [ab]	1135.65 ± 235.56 [a]	1117.79 ± 170.72 [a]	0.041
Chao1	1049.41 ± 225.04 [ab]	911.22 ± 144.92 [b]	997.21 ± 172.15 [ab]	1134.51 ± 235.19 [a]	1115.17 ± 169.25 [a]	0.041
Shannon	5.76 ± 0.48	5.79 ± 0.40	5.95 ± 0.41	6.04 ± 0.43	6.02 ± 0.28	0.316
Shannoneven	0.832 ± 0.042	0.850 ± 0.040	0.864 ± 0.043	0.861 ± 0.035	0.861 ± 0.025	0.237

[1] CON: control group with basal diet; MIC: added compound probiotics (2% of the feed amount) on the basis of CON; MEY: added enzyme and yeast combination preparation (1% of the feed amount) on the basis of MIC; LFC: low-proportion fermented cotton straw (FCS accounted for 14.50% of the diet) group; HFC: high-proportion fermented cotton straw (FCS accounted for 29.00% of the diet) group. Note: In the same row, values with different letter superscripts mean significant differences ($p < 0.05$).

The PCoA analysis based on the Bray–Curtis distance algorithm was conducted to assess the beta-diversity of rumen microbiota (Figure 3). Significant separation of rumen microbiota was observed between different groups of sheep in both stages (p = 0.001).

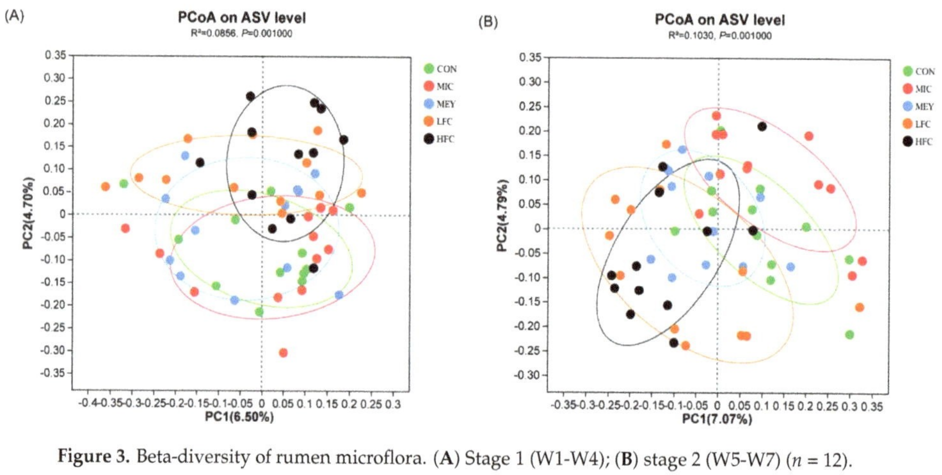

Figure 3. Beta-diversity of rumen microflora. (**A**) Stage 1 (W1-W4); (**B**) stage 2 (W5-W7) (n = 12).

3.6. Effects of Fermented Cotton Straw on the Composition of Rumen Microbiota

In stage 1 (W1–W4), the four most abundant bacterial phyla were Firmicutes, Bacteroidota, Desulfobacterota, and Patescibacteria (Figure 4A). Firmicutes and Bacteroidota exhibited the highest abundance in the CON and HFC groups, respectively. In the treatment groups, the abundance of Firmicutes decreased, while that of Bacteroidota increased. In stage 2, Bacteroidota, Firmicutes, Spirochaetota, and Patescibacteria were the four most abundant bacterial phyla (Figure 4B). Bacteroidota showed the highest abundance in the LFC group.

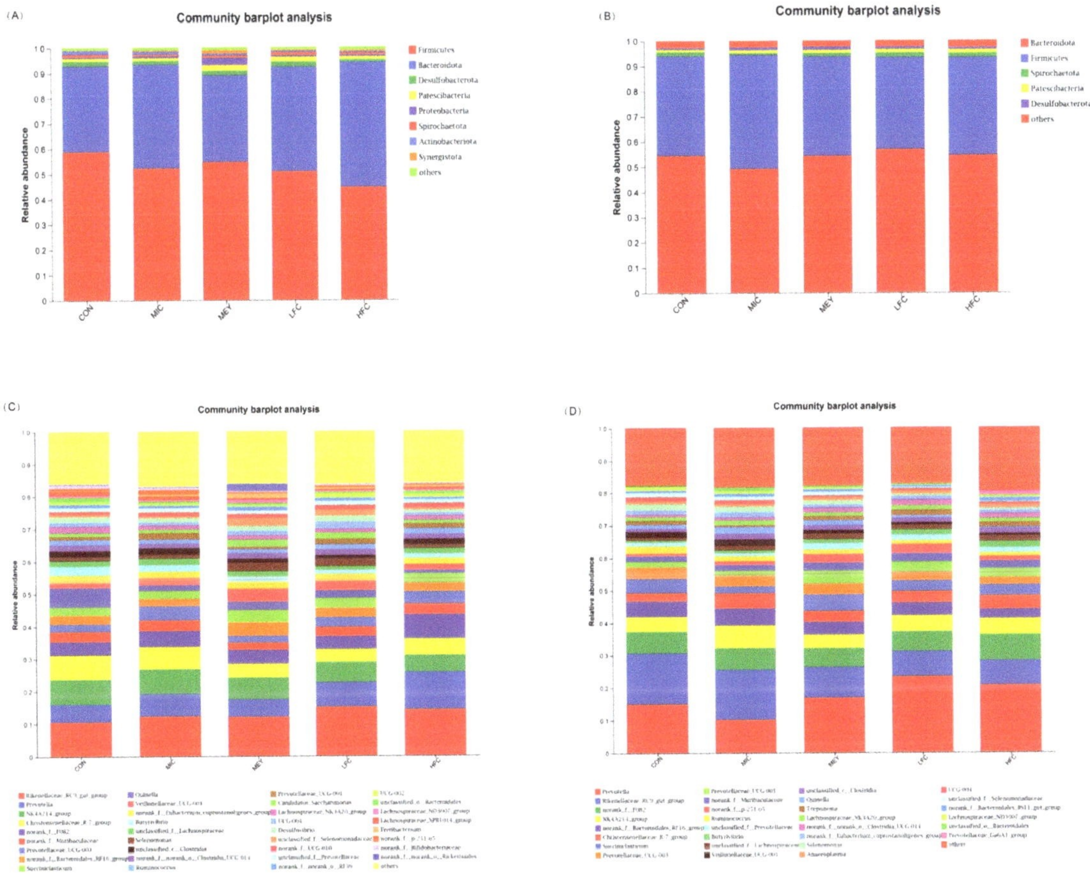

Figure 4. The composition of rumen microflora. (**A**) Stage 1 (W1-W4) for phylum level; (**B**) stage 2 (W5-W7) for phylum level; (**C**) stage 1 (W1-W4) for genus level; (**D**) stage 2 (W5-W7) for genus level (n = 12).

The results in Figure 4C illustrated that the top five most abundant genera were Rikenellaceae_RC9_gut_group, Prevotella, NK4A214_group, Christensenellaceae_R-7_group, and norank_f_F082. Rikenellaceae_RC9_gut_group and Prevotella exhibited the highest abundance in the LFC and HFC groups, respectively. The relative abundance of NK4A214_group and Christensenellaceae_R-7_group decreased in the treatment groups, with the lowest levels observed in the HFC and LFC groups. The composition of rumen microbiota at the genus level in stage 2 (W4–W7) is shown in Figure 4D. The top five genera were Prevotella, Rikenellaceae_RC9_gut_group, norank_f_F082, NK4A214_group, and norank_f_Bacteroidales_RF16_group. The relative abundance of Prevotella increased in all groups except for the MIC group, with the highest abundance in the LFC group. The abundance of Rikenellaceae_RC9_gut_group decreased, with the lowest level observed in the HFC group. As for NK4A214_group, its abundance increased in all groups except for the MEY group, with the MIC group showing the highest level. The abundance of norank_f_Bacteroidales_RF16_group decreased in all groups except for the MIC group.

4. Discussion

4.1. *Effects of Fermented Cotton Straw with Compound Probiotics*

The fermentation trial demonstrated that the crude protein (CP) content of cotton straw significantly increased, while the acid detergent fiber (ADF) and neutral detergent fiber (NDF) contents significantly decreased ($p < 0.05$). Consequently, the nutritional quality of the cotton straw was markedly improved. These findings are consistent with previous studies [21–23]. Biological fermentation has been shown to reduce NDF and ADF levels in straw-based feeds while enhancing protein content, thereby facilitating the efficient utilization of feed resources [24–27]. Additionally, biological fermentation significantly reduced the residues of AFB1, DON, ZEN, FG ($p < 0.05$). Thus, the hygiene and safety of fermented cotton straw were ensured, enabling subsequent feeding trials.

4.2. *Effects of Fermented Cotton Straw on Growth Performance of Sheep*

The indicators of livestock growth performance mainly include body weight (BW), dry matter intake (DMI), average daily gain (ADG), and ratio of feed-to-gain (F/G). Among them, F/G is the important indicator for judging the growth performance of livestock. Based on the effects of fermented cotton straw on the growth performance of sheep, it was demonstrated that fermented cotton straw tends to improve the feed conversion ratio (FCR) of sheep, which is similar to the previous finding [28]. Fermented straw improved the apparent digestibility of livestock; this may be why fermented straw can enhance the growth performance of livestock [29]. For example, lactobacillus-driven feed fermentation has been shown to significantly improve ADG and F/G in pigs. Similarly, feeding livestock with whole-plant corn silage inoculated with lignocellulose-degrading bacteria can significantly increase ADG, DM intake, and F/G. Comparable findings have been observed in the enzyme- and bacteria-driven fermentation of buckwheat straw and alfalfa fed to Tan sheep [11,12,30]. These results align with our experimental findings, demonstrating that fermentation of cotton straw with probiotic complexes can enhance the growth performance of sheep.

4.3. *Effects of Fermented Cotton Straw on Serum Biochemical Parameters of Sheep*

Serum biochemical indicators reflect the health status of animals. TG and TC are used to evaluate lipid utilization in the body, with lower levels indicating better fat utilization. In this study, the TG and TC levels in the treatment group decreased, which is consistent with the findings of Khattab [31]. UREA is an indicator of feed protein utilization and body protein metabolism in sheep [32]. In stage 2, the UREA levels in the CON group were higher, suggesting enhanced utilization of feed protein [33]. Butyric acid is produced during the fermentation of rumen microorganisms, and it undergoes oxidation in rumen epithelial cells to produce beta-hydroxybutyric acid (BHBA). As a biomarker, BHBA plays a critical role in the development of rumen epithelial metabolic function. Moreover, BHBA is a major component of ketone bodies, reflecting ketone body production in the blood [24,34]. In this study, the BHBA content was significantly lower in the treatment group, indicating a lower risk of ketoacidosis in the animals. Acetyl-CoA is a key intermediate linking various metabolic processes, including the tricarboxylic acid cycle, amino acid metabolism, and fatty acid metabolism [35,36]. The somatotropin axis plays a critical role in the regulation of animal growth, primarily involving growth hormone-releasing factor (GRF), growth hormone (GH), and insulin-like growth factor (IGF). GH is the primary hormone responsible for regulating growth and development [37].

The increase in reactive oxygen species (ROS) leads to oxidative stress and cell membrane damage, impairing cell homeostasis, structure, and function. This includes the loss of cell membrane integrity, energy depletion, alterations in signal transduction pathways, and the triggering of apoptosis, among other effects [38]. Antioxidant capacity indicators mainly include T-AOC, SOD, and MDA. T-AOC encompasses various antioxidant enzymes and related biomolecules that can scavenge free radicals in specific organs or organisms, reflecting the functional status of the body's antioxidant system [39]. SOD catalyzes the disproportionation of superoxide ions, removing harmful substances such as free radicals produced during animal metabolism. Thus, it serves as an important indicator of an animal's antioxidant capacity [40]. MDA is a compound formed from the peroxidation of fatty acids and is often used as a marker of ROS. It has toxic effects on cells, primarily by binding to and oxidizing DNA, causing the cross-linking of nucleic acid bases and reacting with other cellular amine groups [41,42]. In this research, MIC, MEY, LFC, and HFC significantly enhanced the antioxidant capacity of sheep ($p < 0.05$), which is consistent with the findings of previous research [43,44].

4.4. Effects of Fermented Cotton Straw on Rumen pH and Microbiota Diversity of Sheep

The normal pH range of rumen fluid is between 5.50 and 7.50. If the pH is too high, the absorption of volatile organic acid (VFAs) by the rumen epithelium can be impaired. Conversely, if the pH is too low, it hinders the fermentation of feed in the rumen [45]. In the present study, the pH of rumen fluid in the test sheep remained within the normal range, indicating that it did not adversely affect fermentation processes.

Rumen microbial diversity includes both alpha diversity and beta diversity. Regarding alpha diversity, indices such as Sobs, Chao1, and ACE reflect the richness of the rumen microbial community. In stage 1, LFC, HFC, MIC, and MEY tended to enhance the richness of the rumen microbiota, which is consistent with the findings of Mamuad [46] and Zhang [47]. However, the community richness of rumen microbiota for MEY and MIC group got lower in stage 2. The Shannon index, which reflects the diversity of the rumen microbiota, showed a tendency to increase in the LFC, HFC, MIC, and MEY groups, consistent with the results reported by Zhang [47] and Rabee [48]. Shannoneven reflects evenness of rumen microbiota. The result illustrated that Shannoneven tended to get higher in each treatment group. Coverage means species coverage of rumen microbiota. With higher coverage level in the treatment group, the probability that the sequence was not detected in the sample of the treatment group is lower, which is similar to the research of Chen [49]. As for beta diversity, fermented cotton straw and probiotics preparation resulted in significant separation of rumen microbiota between different groups.

Bacteroidota and Firmicutes were the dominant bacterial phyla in the rumen microbiota of sheep. Bacteroidota primarily degrade non-fibrous carbohydrates, while Firmicutes are involved in cellulose degradation, which is consistent with the previous finding [11,50,51]. *Prevotella* play a crucial role in the utilization of nutrients such as xylan, pectin, and protein. An increased abundance of *Prevotella* spp. leads to the enhanced expression of associated glycoside hydrolases (GHs) in the rumen. *Prevotella* is recognized for its strong cellulolytic activity and ability to ferment carbohydrates into short-chain fatty acids (SCFAs), playing a crucial role in ruminal protein and carbohydrate metabolism [52]. In the present study, LFC, HFC, and MEY promoted the growth and proliferation of *Prevotella*, thereby increasing the utilization of xylan, pectin, and protein in sheep, consistent with previous studies [12,53,54]. The increased abundance of bacteria associated with protein and carbohydrate utilization may explain the higher feed utilization efficiency in the LFC, HFC, MIC, and MEY groups compared to the CON group, with the F/G ratio in the LFC group being significantly lower than that in the control group ($p < 0.05$). Therefore, fermented cotton straw and probiotic supplementation may improve nutrient utilization in sheep by increasing the abundance of bacteria involved in protein and carbohydrate metabolism, thereby enhancing growth performance.

5. Conclusions

In summary, the diet containing a low proportion of fermented cotton straw (14.50% of total feed) yielded the best short-term fattening effect in sheep (approximately 50 days). Additionally, it improved the utilization of fat and protein, as well as the antioxidant capacity in sheep. Fermented cotton straw enhanced sheep's production performance by increasing the abundance of bacteria involved in the utilization of protein, carbohydrates, and other nutrients, such as *Prevotella*. The fermentation of cotton stalks can efficiently convert agricultural waste into animal feed, promoting sustainable agricultural development.

Author Contributions: Conceptualization, P.W., P.G., Y.Y.; methodology, P.W., M.G., X.L., N.C.; formal analysis, K.Y., X.L., M.G.; writing—original draft preparation, P.W., M.G.; project administration, P.G., Y.Y. All authors have read and agreed to the published version of the manuscript.

Funding: This research was funded by the "Two Areas" Science and Technology Development Program Project (2023LQ02002), the Key Technology R&D Program of Xinjiang Groups (2020AB016), and the China Agriculture Research System (CARS-39-12).

Institutional Review Board Statement: The animal study was approved by the Protocol Management and Review Committee of the Feed Research Institute of Xinjiang Academy of Animal Sciences (Approval No.: 2022017) on 16 July 2022. The study was conducted in accordance with the local legislation and institutional requirements.

Informed Consent Statement: Informed consent was obtained from all subjects involved in the study.

Data Availability Statement: Raw Illumina sequencing data used in this study have been deposited in the Sequence Read Archive (SRA) database of NCBI under Bioproject No. PRJNA964488 (link: https://www.ncbi.nlm.nih.gov/bioproject/PRJNA964488 accessed on 1 May 2023).

Acknowledgments: Thanks to all participants for their advice and support of this study.

Conflicts of Interest: All authors declare no conflicts of interest.

References

1. Baydar, G.; Ciliz, N.; Mammadov, A. Life Cycle Assessment of Cotton Textile Products in Turkey. *Resour. Conserv. Recycl.* **2015**, *104*, 213–223. [CrossRef]
2. Qanmber, G.; Liu, Z.; Li, F.; Yang, Z. Brassinosteroids in Cotton: Orchestrating Fiber Development. *New Phytol.* **2024**, *244*, 1732–1741. [CrossRef] [PubMed]
3. Li, G.; Wang, Y.; Zhu, P.; Zhao, G.; Liu, C.; Zhao, H. Functional Characterization of Laccase Isozyme (PoLcc1) from the Edible Mushroom Pleurotus Ostreatus Involved in Lignin Degradation in Cotton Straw. *Int. J. Mol. Sci.* **2022**, *23*, 13545. [CrossRef] [PubMed]
4. Ren, J.; Yu, P.; Xu, X. Straw Utilization in China—Status and Recommendations. *Sustainability* **2019**, *11*, 1762. [CrossRef]
5. Yang, W.; Li, X.; Zhang, Y. Research Progress and the Development Trend of the Utilization of Crop Straw Biomass Resources in China. *Front. Chem.* **2022**, *10*, 904660. [CrossRef]

6. Datsomor, O.; Yan, Q.; Wang, K.; Mohamed, S.; Opoku-Mensah, L.; Zhao, G.; Miao, L. Effect of Ammoniated and/or Basidiomycete White-Rot Fungi Treatment on Rice Straw Proximate Composition, Cell Wall Component, and In Vitro Rumen Fermentation Characteristics. *Fermentation* **2022**, *8*, 228. [CrossRef]
7. Fan, Z.; Chen, T.; Cai, G.; Huang, X.; Zhong, S.; Li, X.; Zhang, E. Effect of Aspergillus Niger Fermentation on the Metabolites in Corn Stalks. *Fermentation* **2023**, *9*, 50. [CrossRef]
8. Kaur, P.; Bohidar, H.B.; Pfeffer, F.M.; Williams, R.; Agrawal, R. A comparative assessment of biomass pretreatment methods for the sustainable industrial upscaling of rice straw into cellulose. *Cellulose* **2023**, *30*, 4247–4261. [CrossRef]
9. Zhang, Y.X.; Ke, W.C.; Bai, J.; Li, F.H.; Xu, D.M.; Ding, Z.T.; Guo, X.S. The Effect of Pediococcus Acidilactici J17 with High-Antioxidant Activity on Antioxidant, α-tocopherol, β-carotene, Fatty Acids, and Fermentation Profiles of Alfalfa Silage Ensiled at Two Different Dry Matter Contents. *Anim. Feed. Sci. Technol.* **2020**, *268*, 114614. [CrossRef]
10. Yang, G.; Yang, D.; Wang, X.; Cao, W. A Novel Thermostable Cellulase-Producing Bacillus Licheniformis A5 Acts Synergistically with Bacillus Subtilis B2 to Improve Degradation of Chinese Distillers' Grains. *Bioresour. Technol.* **2021**, *325*, 124729. [CrossRef]
11. Jiang, B.; Wang, T.; Zhou, Y.; Li, F. Effects of Enzyme + Bacteria Treatment on Growth Performance, Rumen Bacterial Diversity, KEGG Pathways, and the CAZy Spectrum of Tan Sheep. *Bioengineered* **2020**, *11*, 1221–1232. [CrossRef] [PubMed]
12. Guo, W.; Guo, X.J.; Xu, L.N.; Shao, L.W.; Zhu, B.C.; Liu, H.; Wang, Y.J.; Gao, K.Y. Effect of Whole-Plant Corn Silage Treated with Lignocellulose-Degrading Bacteria on Growth Performance, Rumen Fermentation, and Rumen Microflora in Sheep. *Animal* **2022**, *16*, 100576. [CrossRef]
13. Gao, Y.; Wei, W.; Tian, F.; Li, J.; Wang, Y.; Qi, J.; Xue, S. Corn Straw Total Mix Dietary Supplementation of Bacillus Subtilis-Enhanced Growth Performance of Lambs by Favorably Modulating Rumen Bacterial Microbiome. *Fermentation* **2022**, *9*, 32. [CrossRef]
14. Vataščinová, T.; Pipová, M.; Fraqueza, M.J.R.; Maľa, P.; Dudriková, E.; Drážovská, M.; Lauková, A. Short Communication: Antimicrobial Potential of Lactobacillus Plantarum Strains Isolated from Slovak Raw Sheep Milk Cheeses. *J. Dairy Sci.* **2020**, *103*, 6900–6903. [CrossRef]
15. Abdel-Kareem, M.M.; Rasmey, A.M.; Zohri, A.A. The Action Mechanism and Biocontrol Potentiality of Novel Isolates of Saccharomyces Cerevisiae against the Aflatoxigenic Aspergillus Flavus. *Lett. Appl. Microbiol.* **2019**, *68*, 104–111. [CrossRef] [PubMed]
16. Branà, M.T.; Sergio, L.; Haidukowski, M.; Logrieco, A.F.; Altomare, C. Degradation of Aflatoxin B1 by a Sustainable Enzymatic Extract from Spent Mushroom Substrate of Pleurotus Eryngii. *Toxins* **2020**, *12*, 49. [CrossRef]
17. Zhang, L.; Yang, X.; Huang, R.; Nie, C.; Niu, J.; Chen, C.; Zhang, W. Biodegradation of Free Gossypol by Helicoverpa Armigera Carboxylesterase Expressed in Pichia Pastoris. *Toxins* **2022**, *14*, 816. [CrossRef]
18. AOAC. *Official Methods of Analysis*; Association of Official Analytical Chemists: Washington, DC, USA, 2010.
19. Van Soest, P.J.; Robertson, J.B.; Lewis, B.A. Methods for Dietary Fiber, Neutral Detergent Fiber, and Nonstarch Polysaccharides in Relation to Animal Nutrition. *J. Dairy Sci.* **1991**, *74*, 3583–3597. [CrossRef]
20. Li, J.; Gao, T.; Hao, Z.; Guo, X.; Zhu, B. Anaerobic Solid-State Fermentation with Bacillus Subtilis for Digesting Free Gossypol and Improving Nutritional Quality in Cottonseed Meal. *Front. Nutr.* **2022**, *9*, 1017637. [CrossRef]
21. Khonkhaeng, B.; Cherdthong, A. Improving Nutritive Value of Purple Field Corn Residue and Rice Straw by Culturing with White-Rot Fungi. *J. Fungi* **2020**, *6*, 69. [CrossRef]
22. Qi, X.; Li, Z.; Akami, M.; Mansour, A.; Niu, C. Fermented Crop Straws by Trichoderma Viride and Saccharomyces Cerevisiae Enhanced the Bioconversion Rate of Musca Domestica (Diptera: Muscidae). *Environ. Sci. Pollut. Res. Int.* **2019**, *26*, 29388–29396. [CrossRef] [PubMed]
23. Zhao, X.; Wang, F.; Fang, Y.; Zhou, D.; Wang, S.; Wu, D.; Wang, L.; Zhong, R. High-Potency White-Rot Fungal Strains and Duration of Fermentation to Optimize Corn Straw as Ruminant Feed. *Bioresour. Technol.* **2020**, *312*, 123512. [CrossRef] [PubMed]
24. Elolimy, A.; Vailati-Riboni, M.; Liang, Y.; Loor, J.J. Cellular Mechanisms and Epigenetic Changes: Role of Nutrition in Livestock. *Vet. Clin. N. Am. Food Anim. Pract.* **2019**, *35*, 249–263. [CrossRef] [PubMed]
25. Oskoueian, E.; Jahromi, M.F.; Jafari, S.; Shakeri, M.; Le, H.H.; Ebrahimi, M. Manipulation of Rice Straw Silage Fermentation with Different Types of Lactic Acid Bacteria Inoculant Affects Rumen Microbial Fermentation Characteristics and Methane Production. *Vet. Sci.* **2021**, *8*, 100. [CrossRef]
26. So, S.; Cherdthong, A.; Wanapat, M.; Uriyapongson, S. Fermented Sugarcane Bagasse with Lactobacillus Combined with Cellulase and Molasses Promotes in Vitro Gas Kinetics, Degradability, and Ruminal Fermentation Patterns Compared to Rice Straw. *Anim. Biotechnol.* **2022**, *33*, 116–127. [CrossRef]
27. Zhao, J.; Dong, Z.; Li, J.; Chen, L.; Bai, Y.; Jia, Y.; Shao, T. Effects of Lactic Acid Bacteria and Molasses on Fermentation Dynamics, Structural and Nonstructural Carbohydrate Composition and in Vitro Ruminal Fermentation of Rice Straw Silage. *Asian-Australas. J. Anim. Sci.* **2019**, *32*, 783–791. [CrossRef]

28. So, S.; Cherdthong, A.; Wanapat, M. Growth Performances, Nutrient Digestibility, Ruminal Fermentation and Energy Partition of Thai Native Steers Fed Exclusive Rice Straw and Fermented Sugarcane Bagasse with Lactobacillus, Cellulase and Molasses *J. Anim. Physiol. Anim. Nutr.* **2022**, *106*, 45–54. [CrossRef]
29. Cherdthong, A.; Suntara, C.; Khota, W.; Wanapat, M. Feed Utilization and Rumen Fermentation Characteristics of Thai-Indigenous Beef Cattle Fed Ensiled Rice Straw with Lactobacillus Casei TH14, Molasses, and Cellulase Enzymes. *Livest. Sci.* **2021**, *245*, 104405 [CrossRef]
30. Zhang, D.; Ji, H.; Wang, S.; Liu, Y.; Chen, M.; Liu, H. Lactobacillus-Driven Feed Fermentation Regulates Microbiota Metabolism and Reduces Odor Emission from the Feces of Pigs. *mSystems* **2023**, *8*, e00988-23. [CrossRef]
31. Khattab, I.M.; Abdel-Wahed, A.M.; Khattab, A.S.; Anele, U.Y.; El-Keredy, A.; Zaher, M. Effect of Dietary Probiotics Supplementation on Intake and Production Performance of Ewes Fed *Atriplex* Hay-Based Diet. *Livest. Sci.* **2020**, *237*, 104065. [CrossRef]
32. Schmidely, P.H.; Bahloul, L. Milk Performance and Oxidative Status Responses to Rumen Protected Methionine Supplementation in Genotyped α-S1 Casein Lactating Dairy Goats Fed Two Levels of Metabolizable Protein Diets. *Small Ruminant Res.* **2022**, *209*, 106638. [CrossRef]
33. Li, M.M.; Titgemeyer, E.C.; Hanigan, M.D. A Revised Representation of Urea and Ammonia Nitrogen Recycling and Use in the Molly Cow Model. *J. Dairy Sci.* **2019**, *102*, 5109–5129. [CrossRef] [PubMed]
34. Abdelsattar, M.M.; Vargas-Bello-Pérez, E.; Zhuang, Y.; Fu, Y.; Zhang, N. Impact of Dietary Supplementation of β-Hydroxybutyric Acid on Performance, Nutrient Digestibility, Organ Development and Serum Stress Indicators in Early-Weaned Goat Kids. *Anim. Nutr.* **2022**, *9*, 16–22. [CrossRef] [PubMed]
35. Metallo, C.M.; Gameiro, P.A.; Bell, E.L.; Mattaini, K.R.; Yang, J.; Hiller, K.; Jewell, C.M.; Johnson, Z.R.; Irvine, D.J.; Guarente, L.; et al. Reductive Glutamine Metabolism by IDH1 Mediates Lipogenesis under Hypoxia. *Nature* **2012**, *481*, 380–384. [CrossRef]
36. Sofeo, N.; Hart, J.H.; Butler, B.; Oliver, D.J.; Yandeau-Nelson, M.D.; Nikolau, B.J. Altering the Substrate Specificity of Acetyl-CoA Synthetase by Rational Mutagenesis of the Carboxylate Binding Pocket. *ACS Synth. Biol.* **2019**, *8*, 1325–1336. [CrossRef]
37. Morisset, J. The Early Story of Growth Hormone-Releasing Factor in Rats, Swine, and Cattle. *Can. J. Anim. Sci.* **2022**, *102*, 211–231. [CrossRef]
38. Mameri, A.; Bournine, L.; Mouni, L.; Bensalem, S.; Iguer-Ouada, M. Oxidative Stress as an Underlying Mechanism of Anticancer Drugs Cytotoxicity on Human Red Blood Cells' Membrane. *Toxicol. In Vitro* **2021**, *72*, 105106. [CrossRef]
39. Mao, C.; Xu, Y.; Shi, L.; Guo, S.; Jin, X.; Yan, S.; Shi, B. Effects of Photoperiod Change on Melatonin Secretion, Immune Function and Antioxidant Status of Cashmere Goats. *Animals* **2019**, *9*, 766. [CrossRef]
40. Giorgio, D.; Di Trana, A.; Di Gregorio, P.; Rando, A.; Avondo, M.; Bonanno, A.; Valenti, B.; Di Grigoli, A. Oxidative Status of Goats with Different CSN1S1 Genotypes Fed Ad Libitum with Fresh and Dry Forages. *Antioxidants* **2020**, *9*, 224. [CrossRef]
41. Angrimani, D.S.R.; Silva, R.O.C.; Losano, J.D.A.; Dalmazzo, A.; Tsunoda, R.H.; Perez, E.G.A.; Góes, P.A.A.; Barnabe, V.H.; Nichi, M. Extender Supplementation with Antioxidants Selected after the Evaluation of Sperm Susceptibility to Oxidative Challenges in Goats. *Anim. Biotechnol.* **2019**, *30*, 21–29. [CrossRef]
42. Oldman, A.H.; Martin, D.S.; Feelisch, M.; Grocott, M.P.W.; Cumpstey, A.F. Effects of Perioperative Oxygen Concentration on Oxidative Stress in Adult Surgical Patients: A Systematic Review. *Br. J. Anaesth.* **2021**, *126*, 622–632. [CrossRef] [PubMed]
43. Angulo, M.; Reyes-Becerril, M.; Cepeda-Palacios, R.; Tovar-Ramírez, D.; Esteban, M.Á.; Angulo, C. Probiotic Effects of Marine Debaryomyces Hansenii CBS 8339 on Innate Immune and Antioxidant Parameters in Newborn Goats. *Appl. Microbiol. Biotechnol.* **2019**, *103*, 2339–2352. [CrossRef] [PubMed]
44. Mavrommatis, A.; Mitsiopoulou, C.; Christodoulou, C.; Karabinas, D.; Nenov, V.; Zervas, G.; Tsiplakou, E. Dietary Supplementation of a Live Yeast Product on Dairy Sheep Milk Performance, Oxidative and Immune Status in Peripartum Period. *J. Fungi* **2020**, *6*, 334. [CrossRef] [PubMed]
45. Padilla, L. Impact of pH and Palmitic Acid on Ruminal Fermentation and Microbial Community Composition. Master's Thesis, Utah State University, Logan, UT, USA, 5 May 2022. [CrossRef]
46. Mamuad, L.L.; Kim, S.H.; Biswas, A.A.; Yu, Z.; Cho, K.-K.; Kim, S.-B.; Lee, K.; Lee, S.S. Rumen Fermentation and Microbial Community Composition Influenced by Live Enterococcus Faecium Supplementation. *AMB Express* **2019**, *9*, 123. [CrossRef]
47. Zhang, N.; Wang, L.; Wei, Y. Effects of Bacillus Amyloliquefaciens and Bacillus Pumilus on Rumen and Intestine Morphology and Microbiota in Weanling Jintang Black Goat. *Animals* **2020**, *10*, 1604. [CrossRef]
48. Rabee, A.E.; Younan, B.R.; Kewan, K.Z.; Sabra, E.A.; Lamara, M. Modulation of Rumen Bacterial Community and Feed Utilization in Camel and Sheep Using Combined Supplementation of Live Yeast and Microalgae. *Sci. Rep.* **2022**, *12*, 12990. [CrossRef]
49. Chen, H.; Guo, B.; Yang, M.; Luo, J.; Hu, Y.; Qu, M.; Song, X. Response of Growth Performance, Blood Biochemistry Indices, and Rumen Bacterial Diversity in Lambs to Diets Containing Supplemental Probiotics and Chinese Medicine Polysaccharides. *Front. Vet. Sci.* **2021**, *8*, 681389. [CrossRef]

50. Gharechahi, J.; Vahidi, M.F.; Sharifi, G.; Ariaeenejad, S.; Ding, X.-Z.; Han, J.-L.; Salekdeh, G.H. Lignocellulose Degradation by Rumen Bacterial Communities: New Insights from Metagenome Analyses. *Environ. Res.* **2023**, *229*, 115925. [CrossRef]
51. Guo, Y.; Fan, Z.; Li, M.; Xie, H.; Peng, L.; Yang, C. Effects of Sodium Nitrate and Coated Methionine on Lactation Performance, Rumen Fermentation Characteristics, Amino Acid Metabolism, and Microbial Communities in Lactating Buffaloes. *Microorganisms* **2023**, *11*, 675. [CrossRef]
52. Xue, M.-Y.; Sun, H.-Z.; Wu, X.-H.; Liu, J.-X.; Guan, L.L. Multi-Omics Reveals That the Rumen Microbiome and Its Metabolome Together with the Host Metabolome Contribute to Individualized Dairy Cow Performance. *Microbiome* **2020**, *8*, 64. [CrossRef]
53. Dao, T.-K.; Do, T.-H.; Le, N.-G.; Nguyen, H.-D.; Nguyen, T.-Q.; Le, T.-T.-H.; Truong, N.-H. Understanding the Role of Prevotella Genus in the Digestion of Lignocellulose and Other Substrates in Vietnamese Native Goats' Rumen by Metagenomic Deep Sequencing. *Animals* **2021**, *11*, 3257. [CrossRef] [PubMed]
54. Liu, S.; Shah, A.M.; Yuan, M.; Kang, K.; Wang, Z.; Wang, L.; Xue, B.; Zou, H.; Zhang, X.; Yu, P.; et al. Effects of Dry Yeast Supplementation on Growth Performance, Rumen Fermentation Characteristics, Slaughter Performance and Microbial Communities in Beef Cattle. *Anim. Biotechnol.* **2022**, *33*, 1150–1160. [CrossRef] [PubMed]

fermentation

MDPI

Article

Fermented Feed Promotes Gut Development by Enhancing Intestinal Stem Cell Expansion via Activation of the Wnt/β-Catenin Signaling Pathway

Haozhan Qu, Gengxiu Zan, Haoyan Li, Xiaofan Wang, Jiayi Zhou, Xiuqi Wang and Huichao Yan *

State Key Laboratory of Swine and Poultry Breeding Industry, Guangdong Laboratory for Lingnan Modern Agriculture, Guangdong Provincial Key Laboratory of Animal Nutrition Control, National Engineering Research Center for Breeding Swine Industry, College of Animal Science, South China Agricultural University, Guangzhou 510642, China; q15716489936@163.com (H.Q.); zangengxiu@163.com (G.Z.); lihaoyan1215@126.com (H.L.); xxw033@scau.edu.cn (X.W.); jyzhou@scau.edu.cn (J.Z.); fzyylab@163.com (X.W.)

* Correspondence: yanhc@scau.edu.cn

Abstract: Fermented feed is extensively used in animal production due to their improved palatability and efficient utilization. This study aimed to explore the effects of fermented feed on growth performance and gut health, particularly through the modulation of intestinal stem cells (ISCs). Twenty-four 28-day-old male weaned piglets were randomly assigned into two groups (n = 12): piglets in a control group fed a basal diet, and an experimental group fed a basal diet replaced with 6% fermented mixed feed (FMF) for 28 days. The results indicated that FMF significantly elevated the growth rate, feed intake, and nutrient digestibility of piglets ($p < 0.05$). Furthermore, FMF supplementation increased the jejunal villus height, transepithelial electrical resistance (TEER) values, and the expression of tight junction proteins (ZO-1 and Claudin1) ($p < 0.05$). Immunohistochemistry (IHC) analysis revealed that FMF increased the number of Olfm4$^+$ ISCs and PCNA$^+$ mitotic cells in jejunal crypts, facilitating the differentiation of ISCs into enterocytes (Villin), goblet cells (MUC2), and enteroendocrine cells (CHGA). Conversely, the protein expression of Cleaved Caspase-3 was decreased in the FMF group ($p < 0.05$). Notably, the Wnt/β-catenin signaling pathway, including FZD7 and Active β-catenin, was significantly upregulated in the jejunum after FMF supplementation. Consistent with these findings, intestinal organoids derived from jejunal crypts in the FMF group demonstrated enhanced growth and increased expression of Lgr5, PCNA, KRT20, and β-catenin ($p < 0.05$). These results suggest that fermented feed promotes intestinal development by enhancing ISC proliferation and differentiation via activation of the Wnt/β-catenin signaling pathway.

Keywords: fermented mixed feed; piglet; intestinal stem cell; intestinal development; Wnt/β-catenin signaling pathway

1. Introduction

The modernization of large-scale pig production has intensified the dependence on bulk feedstuffs and the sensitivity of feed prices [1]. The development of deep mining of feed resources and diversified utilization methods represents effective strategies to overcome this challenge [2]. In particular, the rapid advancement of fermentation processes has greatly expanded the utilization of non-traditional raw materials (e.g., grape skins and peanut shells) and enhanced the nutritional quality of conventional feed ingredients [3,4]. Furthermore, beneficial microorganisms present in fermented feeds contribute to maintaining homeostasis and co-metabolism between the gut microbiome and host intestinal epithelial cells, thereby synergistically supporting long-term intestinal development and meat quality [5,6].

The intestinal epithelial development is driven by intestinal stem cells (ISCs), which specifically differentiate into absorptive cells and secretory cells (goblet cells and enteroendocrine cells) to construct the crypt–villus axis structure and perform the digestive-defence functions [7,8]. The high plasticity of ISCs allows epithelium atrophies under persistent deleterious stresses (DON and enterotoxins) and rapid repair in nutrient-rich environments (glutamate, methionine) [9–11]. Moreover, exogenous and endogenous (crypt microenvironment) mitogens, such as R-spondin 1 and Wnts, stimulate ISC division to facilitate epithelial tissue regeneration by amplifying Wnt/β-catenin signaling [12–14]. Consequently, the precise regulation of ISC proliferation and differentiation is beneficial to intestinal development and repair after injury [15].

Nutrients such as peptides, fats, and carbohydrates in feeds or produced by microbial fermentation of feeds can regulate the regeneration and differentiation of ISCs [16]. Additionally, beneficial bacteria promote intestinal development by producing short-chain fatty acids that serve as an energy source for intestinal epithelial cells [17,18]. This process contributes to the overall health and functionality of the digestive tract.

To investigate the benefits of feed fermentation on animal health, we used models of piglets and demonstrated that fermented mixed feed (FMF) supplementation enhances Wnt/β-catenin signaling and accelerates ISC expansion to promote gut development. This finding provides valuable insights into potential applications of FMF, offering a theoretical basis for its use as a nutritional supplement to support healthy digestive function in pigs.

2. Materials and Methods

2.1. Fermented Mixed Feed Preparation and Chemical Analysis

The preparation method of FMF was based on previous studies [19]. The *Lactobacillus johnsonii*, *Bacillus subtilis*, and *Enterococcus faecium* used in this study were all provided by Kebang Feedstuff Technology Co., Ltd. (Guangzhou, China). The substrate was composed of 40% corn, 35% soybean meal, and 25% bran. After steam sterilization, sterile water was added and continuously stirred to reach a total moisture content of 50%. An amount of 1 kg of *Lactobacillus johnsonii* powder (1×10^8 CFU/g), 1 kg of *Bacillus subtilis* powder (1×10^8 CFU/g), and 1 kg of *Enterococcus faecalis* (3×10^8 CFU/g) were inoculated into 1000 kg of the wet mixed substrate. Subsequently, the wet mixture was transferred into plastic bags with one-way valves, sealed, and fermented at 35 °C for 72 h. The nutritional components of the diets were formulated to meet the NRC (2012) (Table S1). Nutrient requirements are presented in Table 1.

Table 1. Nutrient levels (dry matter basis).

Nutrient Levels	CON	FMF
Moisture (%)	9.19	9.31
Dry matter (%)	90.81	90.69
Crude protein (%)	19.98	20.78
Ether extract (%)	5.09	5.48
Crude fiber (%)	2.49	2.17

CON = basal diet; FMF = fermented feed replacing 6% of the basal diet. The contents of Dry matter (DM), Crude protein (CP), Ether extract (EE), and Crude fiber (CF) are all measured values.

2.2. Animal Experimental Design

Twenty-four weaned piglets (Duroc × Landrace × Large White, body weight = 13.1 ± 0.23 kg) were randomly allocated into the control (CON) group and fermented mixed feed

(FMF) group (n = 12 piglets per group), and fed with a basal diet and a fermented mixed feed (containing 6% fermentation product) for 28 days, respectively. The basic diet and fermented feed were provided by Kebang Feedstuff Technology Co., Ltd. (Guangzhou, China), and the basic diet formulation is shown in Table 2. After the feeding experiment, the pigs were anesthetized with an intravenous injection of sodium pentobarbital (200 mg/kg BW) and slaughtered. Then, the jejunal tissues were collected after washing with phosphate-buffered saline (PBS) for immunohistochemistry (IHC), transmission electron microscopy (TEM), and intestinal organoid (IO) culture. All experiments were approved by the animal ethics committee of South China Agricultural University (Guangzhou, China).

Table 2. Treatments applied in the study.

Items	Basal Diet
Corn	34.00
Soybean meal	23.00
Flour	10.00
Puffed corn	20.00
Corn gluten meal	8.00
Soybean oil	1.50
Calcium hydrogen phosphate	1.00
Sodium chloride	0.80
L-lysine hydrochloride	0.60
DL-methionine	0.03
L-threonine	0.02
L-tryptophan	0.05
Premix	1.00
Total	100.00

Premixes provided for each kilogram: Vitamin A 10,500 IU, Vitamin D_3 2100 IU, Vitamin K_3 3.5 mg, Vitamin B_1 2.24 mg, Vitamin B_2 9.80 mg, Vitamin B_6 2.80 mg, Vitamin B_{12} 0.05 mg, Folic Acid 1.04 mg, Niacin 35 mg, Cu 15 mg, Fe 101 mg, Manganese 42 mg, Zn 102 mg, I 0.60 mg, Se 0.30 mg.

2.3. Determination of Feed Nutrient Digestibility

Proximate nutrient digestibility was assessed using the exogenous indicator method (TiO_2) [20]. Specifically, in the final week of the trial, 0.3% TiO_2 was incorporated into the diet. Piglet fecal samples were collected and treated with 10% hydrochloric acid to stabilize the nitrogen content. The determination of TiO_2 content in both feed and feces samples was conducted at 420 nm using a UV-Vis spectrophotometer (Multiskan SkyHigh, Waltham, MA, USA) following Chinese National Standard GB 5009.246-2016 [21]. Furthermore, after drying at 65 °C for 48 h, the feed and feces samples were crushed through a 1 mm sieve, homogenized, and analyzed for dry matter (DM, GB/T 6435-2014) [22], crude protein (CP, GB/T 6432-2018) [23], ether extract (EE, GB/T 6433-2006) [24], and crude fiber (CF, GB/T 6434-2022) [25] content to calculate their digestibility {apparent digestibility = [(total amount of the nutrient ingested in the feed-total amount of the nutrient in the feces)/total amount of the nutrient ingested in the feed] × 100%}.

2.4. Antibodies

Antibodies against Active β-catenin (#19807, 1:200 [IHC]), KRT20 (#13063, 1:200 [IHC],1:2000 [WES]), and Olfm4 (#14369, 1:200 [IHC]) were procured from Cell Signaling Technology (Boston, MA, USA). Antibodies against Frizzled7 (FZD7, #bs-5125R, 1:200 [IHC]) and Chromogranin (CHGA, #bs-0539R, 1:200 [IHC]) were obtained from Bioss (Beijing, China). Villin (#sc58897, 1:200 [IHC]) antibodies were procured from Santa Cruz Biotechnology (Dallas, TX, USA). Antibodies against antibodies Mucin 2 (MUC2, #A14659, 1:200 [IHC]) were obtained from ABclonal (Wuhan, China); ZO-1 (#511417, 1:200

[IHC]), Claudin1 (#343203, 1:200 [IHC]), Cleaved Caspase-3 (C-Caspase-3, #341034, 1:200 [IHC]), β-catenin (#201328, 1:2000 [WES]), Lgr5 (#R380991, 1:2000 [WES]), PCNA (#200947, 1:200 [IHC],1:2000 [WES]), as well as Anti-rabbit IgG (#511203) and Anti-mouse IgG (#511103) secondary antibodies, were obtained from Zen BioScience (Chengdu, China).

2.5. Hematoxylin and Eosin (H&E) Staining

After one week of fixation in formaldehyde, the intestinal tissue samples were trimmed into slices and placed in embedding cassettes. The samples underwent gradient dehydration using ethanol, followed by clearing in xylene for 40 min. Once cleared, the tissues were immersed in paraffin solution at 60 °C for infiltration. After sufficient infiltration, the intestinal tissues were further trimmed and embedded in paraffin blocks. The blocks were left to cool and solidify, forming wax blocks. These blocks were then sectioned into 5 μm slices (with three non-continuous slices prepared for each group) and dried at 42 °C for 8 h to prepare paraffin sections. Once dried, the wax sections underwent de-waxing with xylene and hydration with a series of graded ethanol solutions. The sections were then incubated with hematoxylin solution for 8 min and eosin solution for 2 min, with water washes performed after each incubation step to remove excess dye. Following this, the sections were dehydrated through graded ethanol, cleared in xylene, and mounted with neutral resin. The morphological structure of the small intestinal crypt–villus axis was observed under an inverted fluorescence microscope (Ti2-U, Nikon, Tokyo, Japan). Images were captured, and the villus height, crypt depth, and villus-to-crypt ratio were quantified using ImageJ software (version 1.8.0 112, National Institutes of Health, Bethesda, MD, USA).

2.6. IHC Staining

After the paraffin sections were dried, they underwent de-waxing with xylene and hydration through a series of graded ethanol solutions. The sections were then placed in antigen retrieval solution for the antigen retrieval procedure. After retrieval, the samples were washed and incubated at room temperature with bovine serum albumin (BSA) for 2 h to block nonspecific binding sites. Following blocking, the sections were incubated overnight at 4 °C with primary antibodies (ZO-1, Claudin1, Olfm4, PCNA, C-Caspase-3, KRT20, Villin, MUC2, CHGA, FZD7, and Active β-catenin), which were diluted 1:200 in antibody dilution buffer. After washing with PBS, the sections were incubated with fluorescent secondary antibodies (diluted 1:200 in antibody dilution buffer) at room temperature for 2 h in the dark. Following another PBS wash to remove excess secondary antibody, the sections were stained with 4′,6-diamidino-2-phenylindole (DAPI) for nuclear staining. The DAPI solution (10 μg/mL) was applied for 10 min, and the sections were washed three times with PBS. Finally, the stained sections were observed and documented under a fluorescence inverted microscope (Ti2-U, Nikon, Tokyo, Japan), and the fluorescence intensity of the target proteins was quantified with ImageJ software (version 1.8.0 112, National Institutes of Health, Bethesda, MD, USA).

2.7. TEM

The jejunal tissues were collected and fixed in 2.5% glutaraldehyde for 24 h and 1% osmium acid for 1 h. Then, the samples were dehydrated in gradient alcohol and anhydrous acetone, embedded in resin, and made into ultrathin sections (Talos F200S, Thermo Fisher Scientific, Waltham, MA, USA).

2.8. Transepithelial Electrical Resistance (TEER)

As previously described, jejunal samples were sectioned into 3×1 cm^2 segments and equilibrated in Krebs–Ringer buffer for 10 min. Subsequently, the samples were fixed between the two halves of an Ussing Chamber (C4300/VCC MC6 Plus, Beijing KingTech Technology Co., Ltd., Beijing, China), representing the apical and basolateral sides. The system was maintained at 37 °C with a water jacket and aerated with carbogen gas (95% O_2, 5% CO_2). Following a 30 min stabilization period, the buffers were refreshed, and the experiments commenced. The TEER values were recorded through three successive readings and reported as Ω (resistance) × cm^2 (monolayer surface area) [26].

2.9. *Intestinal Organoid Culture*

The jejunal tissues in the CON and FMF groups were longitudinally dissected, sectioned into small pieces, and incubated with cold Dulbecco's phosphate-buffered saline (DPBS) containing 30 mM of ethylenediaminetetraacetic acid disodium salt to harvest crypts. Then, the crypts were placed in 48-well plates (Jet Bio-Filtration Co., Ltd., Guangzhou, China) and cultivated into IOs using previously established culture systems [27]. The growth advantages of IOs were assessed based on the IO forming efficiency (the ratio of total IOs to seeded crypts), surface area, budding efficiency (the ratio of budding IOs to total IOs), and branching coefficient (ratio of budding IOs in various classes to total IOs).

2.10. *Protein Expression Detection (WES)*

The IOs were cleaned with cold PBS and lysed in RIPA buffer to obtain protein samples. Subsequently, the protein samples were mixed with a 5× fluorescent master mix after determining the protein concentration using the BCA method and heated at 95 °C for 5 min. The samples, blocking reagent, wash buffer, primary antibodies, secondary antibodies, and chemiluminescent substrate were carefully dispensed into the designated wells of the microplate according to the manufacturer's instructions. The plate was then loaded into the instrument, and the protein samples were drawn into individual capillaries. Automatic protein separation was performed automatically using default settings. Data analysis was carried out using Compass software 4.0 (ProteinSimple, San Jose, CA, USA).

2.11. *Statistical Analysis*

The data are presented as the mean ± SEM. The data were analyzed using SPSS software version 20.0 (SPSS Inc., Chicago, IL, USA). Student's *t*-test was used for statistical analysis. *p*-values are represented as * $p < 0.05$ and ** $p < 0.01$.

3. Results

3.1. *Ingredients and Nutrient Levels of Experimental Diets (%, As-Fed Basis)*

The composition of the basic diet is provided in Table 1. As shown in Table 2, compared with the basal diet, the fermented alternative feed exhibits higher levels of crude protein (20.78% vs. 19.98%) and ether extract (5.48% vs. 5.09%), as well as lower crude fiber content (2.17% vs. 2.49%), indicating that fermentation contributes to the enhancement of the nutritional quality of the feed.

3.2. *FMF Improves the Growth Performance of Piglets*

The growth performance of piglets was evaluated throughout the experimental period, and it was found that the feed intake was significantly higher in the FMF group than in the CON group from day 19 onwards ($p < 0.05$; Figure 1A). From day 22 onwards, the FMF group had consistently higher body weight than the CON group ($p < 0.05$; Figure 1B). The average daily feed intake (ADFI) and average daily gain (ADG) of the FMF group were significantly higher than those of the CON group ($p < 0.05$; Figure 1C), while the Feed to Gain Ratio (F:G) in the FMF group was significantly lower than that in the CON group ($p < 0.05$; Figure 1D).

Additionally, the exogenous marker method indicated that FMF significantly improved the apparent digestibility of DM, CP, EE, and CF ($p < 0.01$; Figure 1E), suggesting that FMF could enhance the utilization of nutrients in the intestine and promote piglet growth.

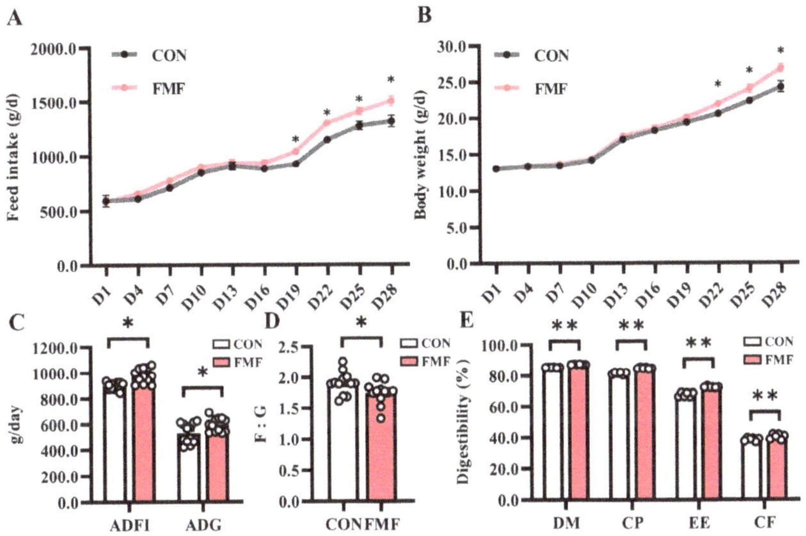

Figure 1. FMF increases the growth performance of piglets. (**A**) Dynamics of piglet feed intake during the experiment (n = 12). (**B**) Dynamics of piglet body weight during the experiment (n = 12). (**C**) Average daily feed intake (ADFI) and average daily gain (ADG) (n = 12). (**D**) Feed to Gain Ration (F: G) (n = 12). (**E**) Apparent digestibility of DM, CP, EE and CF (n = 6). Data represent mean ± SEM; * $p < 0.05$; ** $p < 0.01$; Student's t-test.

3.3. FMF Promotes Intestinal Epithelial Development and Barrier Function

Compared with the CON group, the jejunal mass and the mucosal mass of the jejunum and ileum in the FMF group were significantly increased ($p < 0.05$; Figure 2A,B). The H&E staining showed that FMF improved the villus height and ratio of villus to crypt in the jejunum ($p < 0.05$; Figure 2C–E). The TEM showed that FMF-treated jejunal crypt cells were arranged in an orderly manner (Figure 2F). The Ussing Chamber assay showed that the TEER values significantly increased after fermented feed supplementation ($p < 0.01$; Figure 2G). Furthermore, the data of IHC showed that the protein expression of ZO-1 and Claudin1 in the jejunal tissues was increased by FMF ($p < 0.05$; Figure 2H–J). These results indicate that FMF may effectively support jejunal development and improve its barrier function in piglets.

3.4. FMF Enhances ISC Division and Differentiation

Considering that intestinal epithelial development depends on ISC proliferation and differentiation, we evaluated the characteristics of ISCs. Following FMF administration, an increase in Olfm4 (ISCs marker; Figure 3A,B) fluorescence signal was observed in the jejunal crypts. Concurrently, there was a consistent upregulation trend in the expression levels of proliferative cell marker (PCNA; Figure 3A,C), differentiated cell marker (KRT20; Figure 3A,D), absorptive cell marker (Villin; Figure 3A,E), goblet cell marker (MUC2; Figure 3A,F), and enteroendocrine cell marker (CHGA; Figure 3A,G). Conversely, a significant reduction in the apoptotic marker C-Caspase-3 was observed in the villus ($p < 0.05$, Figure 3A,H). These results substantiate that FMF enhances ISC division and differentiation.

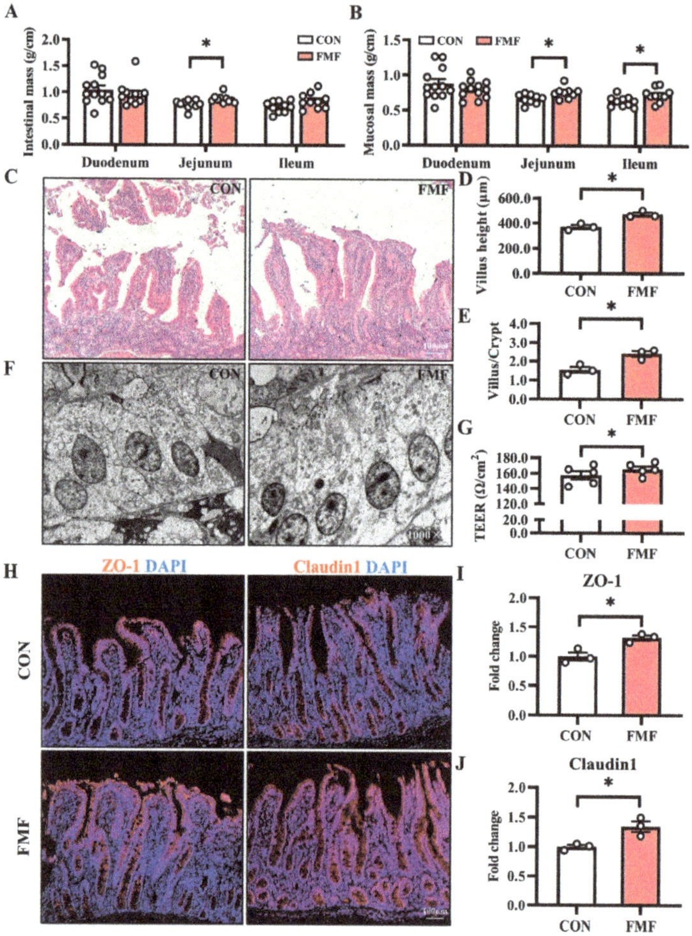

Figure 2. FMF promotes intestinal epithelial development and barrier function. (**A**) Intestinal weight (n = 12). (**B**) Mucosal weight (n = 12). (**C**) H&E staining analyzed the morphological structure of jejunum (100×). (**D**) Villus height (n = 3). (**E**) Ratio of villus height to crypt depth of the jejunum (n = 3). (**F**) Representative images of TEM in the jejunum of piglets (1000×). (**G**) TEER of the jejunum (n = 5). (**H**) Images of immunohistochemical staining with the ZO-1 and Claudin1 antibody in the jejunum (100×). (**I**,**J**) Statistical analysis of the fluorescence intensity of ZO-1 (**I**) and Claudin1 (**J**) (n = 3). Data represent mean ± SEM; * $p < 0.05$; Student's t-test.

3.5. FMF Promotes ISC Expansion Ex Vivo

Furthermore, piglet jejunal crypts isolated from the CON and FMF groups were cultured ex vivo to assess IO growth advantages ($p < 0.05$; Figure 4A). These results showed that IO forming efficiency ($p < 0.05$; Figure 4B) and budding efficiency ($p < 0.05$; Figure 4C) were elevated in the FMF group, while the surface area of IOs did not show a significant change (Figure 4D). Meanwhile, the proportion of IOs with more than 5 buds in the FMF group was significantly higher than that in the CON group ($p < 0.05$; Figure 4E). WES showed that FMF increased the expression of Lgr5 (an ISC marker), PCNA, and KRT20, as well as β-catenin in IOs ($p < 0.05$, Figure 4F,G). Collectively, these results imply that FMF promotes the ISC expansion ex vivo and enhances the growth advantage of IOs.

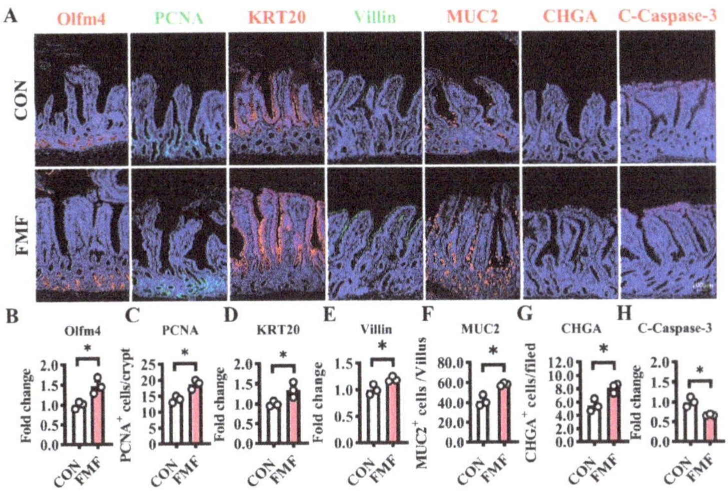

Figure 3. FMF enhances intestinal stem cell (ISC) division and differentiation. (**A**) Images of IHC staining with the Olfm4, PCNA, KRT20, Villin, MUC2, CHGA, and C-Caspase-3 antibody in the jejunum (100×). (**B–H**) Statistical analysis of the fluorescence intensity or positive cells of Olfm4 (**B**), PCNA (**C**), KRT20 (**D**), Villin (**E**), MUC2 (**F**), CHGA (**G**) and C-Caspase-3 (**H**) (n = 3). Data represent mean ± SEM; * $p < 0.05$; Student's t-test.

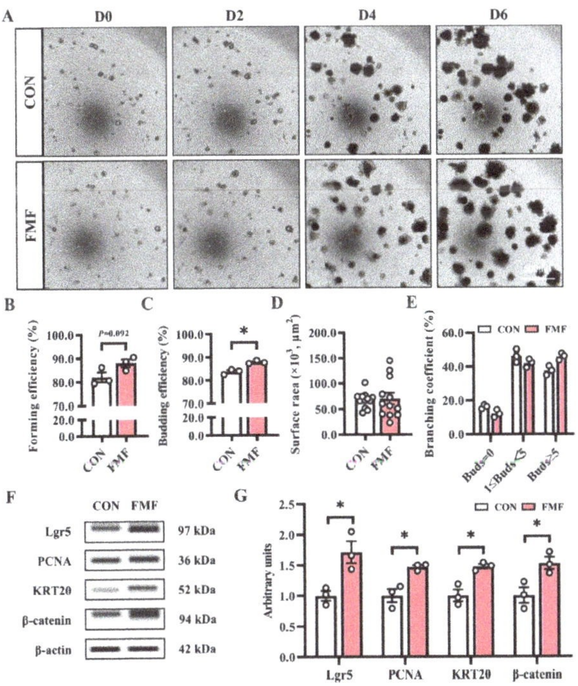

Figure 4. FMF promotes ISC expansion ex vivo. (**A**) Representative images of IOs cultured from crypt cells from each group (40×). (**B**) IOs forming efficiency (n = 3). (**C**) IOs budding efficiency (n = 3). (**D**) IOs surface area (n = 12). (**E**) The results of the statistical analysis of IOs bud number and branching coefficient of IOs (n = 3). (**F,G**) The expressions of Lgr5, PCNA, KRT20, and β-catenin in IOs (n = 3). Data represent mean ± SEM; * $p < 0.05$; Student's t-test.

3.6. FMF Activates the Wnt/β-Catenin Signaling Pathway in the Jejunum

The Wnt/β-catenin signaling is the controller of ISC-driven intestinal epithelial renewal. Our results showed that supplementation with FMF heightened the fluorescence signal intensity ($p < 0.05$, Figure 5A–C) of the Wnt membrane receptor FZD7 and effector Active β-catenin, indicating that FMF amplifies the Wnt/β-catenin signaling pathway.

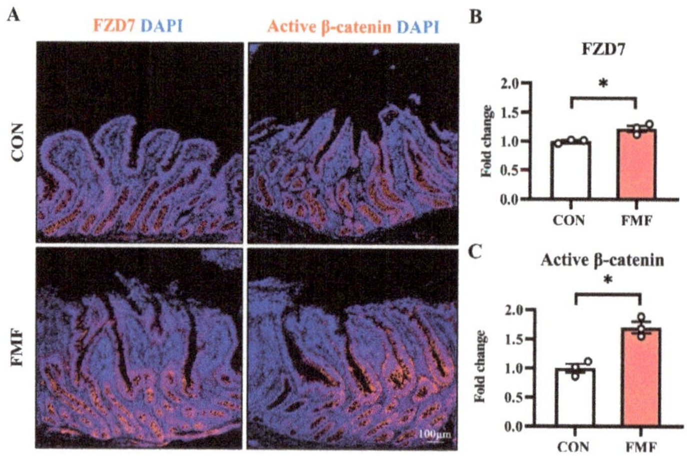

Figure 5. FMF activates the Wnt/β-catenin signaling pathway in the jejunum. (**A**) Images of IHC staining with the FZD7 and Active β-catenin antibody in the jejunum (100×). (**B**,**C**) Statistical analysis of the fluorescence intensity of FZD7 (**B**) and Active β-catenin (**C**) (n = 3). Data represent mean ± SEM; * $p < 0.05$; Student's t-test.

4. Discussion

The rapid development of animal husbandry has met the increasing demand for meat products but has also exacerbated the pressure of feed resource shortage [28]. The introduction of fermentation technology in the production process has improved the structure of feed and optimized its nutritional composition [29]. Extensive data demonstrate that the use of fermented feed can enhance the digestibility of nutrients and promote animal growth [30]. However, the precise regulation of intestinal development by fermented feed as a complex factor remains unclear [31]. In this study, we found that fermented feed could accelerate the ISC proliferation and differentiation into absorptive and secretory cells, thereby enhancing the integrity of intestinal epithelial structure and barrier function using weaned piglets as a model. Mechanistically, the Wnt/β-catenin signaling pathway plays a crucial role in regulating ISC function, providing a theoretical basis for the precise application of fermented feed [32].

Fermented feed is a biologically enriched feed produced by fermenting feed ingredients (mainly including straw, bran, corn, and dregs) using microorganisms and converting them into various high-quality nutrients such as polysaccharides, proteins, and fats [33]. Our data show that 6% fermented feed significantly increases feed intake, promotes the digestion of nutrients in the feed, and accelerates the growth of piglets. In an investigation of microbial dynamics during the fermentation of animal feed, it was observed that the population of lactobacilli began to increase significantly from the first day of fermentation; by the second day, this population stabilized, coinciding with substantial lactic acid production which led to a decrease in feed pH. As the pH declined, the populations of Enterobacteriaceae and mold rapidly diminished to lower levels. On the third day, the feed pH stabilized between 4.0 and 4.2 [34]. Shi et al. demonstrated that the addition of *Bacillus subtilis* and *Enterococcus faecium* to corn for fermentation resulted in a reduction of approximately 30% in the populations of Enterobacteriaceae and mold in the feed [35]. These changes in the physicochemical properties not only enhance the nutritional value of the feed but also improve the efficiency of intestinal microbiota utilization.

For the host, adding fermented feed to broiler feed can extend the length of the jejunum and ileum and effectively eliminate harmful intestinal bacteria [36,37]. Meanwhile, fermented feed is able to activate intestinal immune function and reduce mucosal damage in weaned piglets [38]. The modulation of intestinal microbiota diversity represents a critical mechanism underlying the beneficial effects of fermented feed on piglet intestinal health. Previous research has found that administering fermented wheat feed significantly enhances the diversity of the small intestinal microbiota in piglets, with a notable increase in the abundance of beneficial bacteria such as Lactobacillus and Clostridium butyricum. These alterations in microbial composition can suppress pathogen growth through competition for ecological niches or by modulating the intestinal pH environment, thereby effectively safeguarding intestinal health [39,40]. In the present study, FMF promoted jejunal development by increasing villus height and enhancing tight junctions. Notably, TEM observation reveals that the crypt cells in the FMF group exhibit a more orderly arrangement, which is more conducive to consolidating the ISC niche.

Lgr5-labeled ISCs located at the base of crypts are essential for intestinal epithelial regeneration [41]. These ISCs continuously proliferate and migrate towards the top of the villus, undergoing differentiation into various cell types such as enterocytes, goblet cells, enteroendocrine cells, Paneth cells, and other specialized intestinal cells [42,43]. This dynamic process supported by adequate nutrition ensures the maintenance of intestinal function through cellular turnover [44]. For example, l-glutamate can facilitate ISC renewal in piglets and IOs [10]. Our research also found that supplementation of FMF significantly enhances ISC expansion, which might be attributed to the nutrient-rich composition of the fermented feed [45–47]. Previous studies have demonstrated that microbial metabolites can affect ISC function by regulating mucus, even though protective mucus-secreting cells within crypts serve to prevent microbial invasion [48,49]. Therefore, we hypothesize that the microorganisms present in FMF may play a crucial role in activating ISCs.

The Wnt/β-catenin signaling pathway plays a crucial role in the ISC niche [50]. When ISCs are damaged by stressors such as radiation or toxins, Wnt/β-catenin activation via Wnt enhancer RSPO1 can accelerate ISC turnover and promote epithelial remodeling [51]. In this study, we found that fermented feed enhanced Wnt/β-catenin signaling activity in the piglet jejunum and IOs. However, as a mixture, how fermented feed influences Wnt/β-catenin signaling activation remains unclear. The Wnt signaling pathway, particularly the membrane receptor FZD7, serves as an integrator for various nutrient and microbial metabolite signals [52,53]. Lactic acid and short-chain fatty acids from microbial metabolism have the potential to boost Wnt/β-catenin signaling [54]. Thus, the activation of Wnt/β-catenin signaling by fermented feeds may involve a coordinated process with multiple components. Despite the specific components of fermented feed being difficult to ascertain, adjusting the fermentation process based on changes in the Wnt/β-catenin signaling pathway could potentially enhance its inherent efficacy.

5. Conclusions

Overall, our findings demonstrate that fermented feed stimulates the expansion of FZD7/β-catenin-responsive ISCs, increases the number of absorptive and secretory cells, enhances intestinal barrier function, and promotes piglet development (Figure 6).

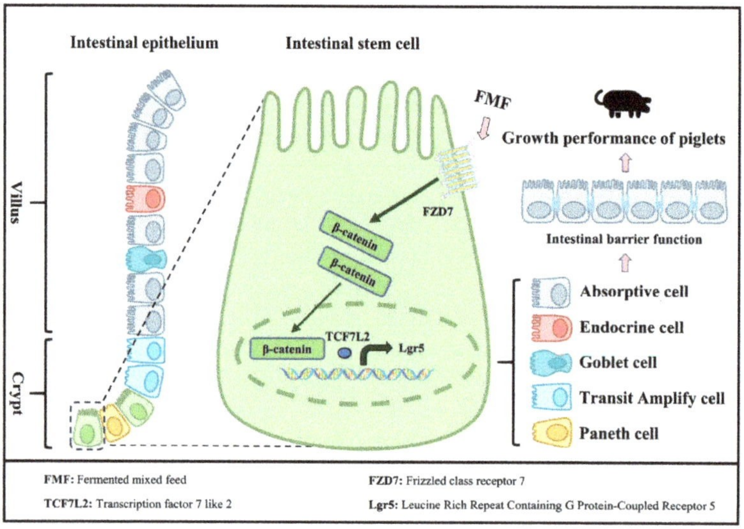

Figure 6. Mechanism of fermented feed promoting intestinal epithelial development driven by ISCs. Fermented feed enhances the activity of ISCs in piglets by activating the Wnt/β-catenin signaling pathway.

Supplementary Materials: The following supporting information can be downloaded at: https://www.mdpi.com/article/10.3390/fermentation11020052/s1, Table S1. NRC (2012) pig nutrition standards (11–25kg).

Author Contributions: H.Q. contributed to the conception and design, data analysis and interpretation, and manuscript writing. G.Z. and H.L. contributed to the collection and assembly of data. J.Z. and X.W. (Xiaofan Wang) contributed to the manuscript editing. H.Y. and X.W. (Xiuqi Wang) contributed to the conception and design, and final approval of the manuscript. All authors have read and agreed to the published version of the manuscript.

Funding: This work was supported by the National Key Research and Development Program of China (2023YFE0124400), the Science and Technology Planning Project of Guangzhou (2024B03J1267), and the Technology Planning Projects of Guangdong Province, China (2022A0505050064).

Institutional Review Board Statement: All animal experiments were performed following the guidelines of the South China Agricultural University Animal Care and Use Committee (12 November 2022; permit no. SCAU2022F416).

Informed Consent Statement: Not applicable.

Data Availability Statement: The data that support the findings of this study are available from the corresponding author upon reasonable request. The data are not publicly available due to privacy.

Conflicts of Interest: The authors declare no conflicts of interest.

References

1. Rauw, W.M.; Rydhmer, L.; Kyriazakis, I.; Øverland, M.; Gilbert, H.; Dekkers, J.C.; Hermesch, S.; Bouquet, A.; Gómez Izquierdo, E.; Louveau, I.; et al. Prospects for sustainability of pig production in relation to climate change and novel feed resources. *J. Sci. Food Agric.* **2020**, *100*, 3575–3586. [CrossRef]
2. Surendra, K.C.; Angelidaki, I.; Khanal, S.K. Bioconversion of waste-to-resources (BWR-2021): Valorization of industrial and agro-wastes to fuel, feed, fertilizer, and biobased products. *Bioresour. Technol.* **2022**, *347*, 126739. [CrossRef]

3. Nardoia, M.; Romero, C.; Brenes, A.; Arija, I.; Viveros, A.; Ruiz-Capillas, C.; Chamorro, S. Addition of fermented and unfermented grape skin in broilers' diets: Effect on digestion, growth performance, intestinal microbiota and oxidative stability of meat. *Animal* **2020**, *14*, 1371–1381. [CrossRef] [PubMed]
4. Liu, J.; Wang, Z.; Li, H.; Hu, C.; Raymer, P.; Huang, Q. Effect of solid state fermentation of peanut shell on its dye adsorption performance. *Bioresour. Technol.* **2018**, *249*, 307–314. [CrossRef] [PubMed]
5. Shi, N.; Li, N.; Duan, X.; Niu, H. Interaction between the gut microbiome and mucosal immune system. *Mil. Med. Res.* **2017**, *4*, 14. [CrossRef] [PubMed]
6. Wen, C.; Wang, Q.; Gu, S.; Jin, J.; Yang, N. Emerging perspectives in the gut-muscle axis: The gut microbiota and its metabolites as important modulators of meat quality. *Microb. Biotechnol.* **2024**, *17*, e14361. [CrossRef]
7. Barker, N. Adult intestinal stem cells: Critical drivers of epithelial homeostasis and regeneration. *Nat. Rev. Mol. Cell Biol.* **2014**, *15*, 19–33. [CrossRef]
8. Bankaitis, E.D.; Ha, A.; Kuo, C.J.; Magness, S.T. Reserve Stem Cells in Intestinal Homeostasis and Injury. *Gastroenterology* **2018**, *155*, 1348–1361. [CrossRef]
9. Zhou, J.Y.; Lin, H.L.; Qin, Y.C.; Li, X.G.; Gao, C.Q.; Yan, H.C.; Wang, X.Q. l-Carnosine Protects Against Deoxynivalenol-Induced Oxidative Stress in Intestinal Stem Cells by Regulating the Keap1/Nrf2 Signaling Pathway. *Mol. Nutr. Food Res.* **2021**, *65*, e2100406. [CrossRef] [PubMed]
10. Qin, Y.C.; Zhou, J.Y.; Zhu, M.; Zan, G.X.; Gao, C.Q.; Yan, H.C.; Li, X.G.; Wang, X.Q. L-glutamate requires β-catenin signalling through Frizzled7 to stimulate porcine intestinal stem cell expansion. *Cell. Mol. Life Sci.* **2022**, *79*, 523. [CrossRef]
11. Zhu, C.; Liang, S.; Zan, G.; Wang, X.; Gao, C.; Yan, H.; Wang, X.; Zhou, J. Selenomethionine Alleviates DON-Induced Oxidative Stress via Modulating Keap1/Nrf2 Signaling in the Small Intestinal Epithelium. *J. Agric. Food Chem.* **2023**, *71*, 895–904. [CrossRef]
12. Zhou, J.Y.; Zan, G.X.; Zhu, Q.J.; Gao, C.Q.; Yan, H.C.; Wang, X.Q. Recombinant Porcine R-Spondin 1 Facilitates Intestinal Stem Cell Expansion along the Crypt-Villus Axis through Potentiating Wnt/β-Catenin Signaling in Homeostasis and Deoxynivalenol Injury. *J. Agric. Food Chem.* **2022**, *70*, 10644–10653. [CrossRef]
13. Barker, N. The canonical Wnt/beta-catenin signalling pathway. *Methods Mol. Biol.* **2008**, *468*, 5–15. [CrossRef] [PubMed]
14. Zou, W.Y.; Blutt, S.E.; Zeng, X.L.; Chen, M.S.; Lo, Y.H.; Castillo-Azofeifa, D.; Klein, O.D.; Shroyer, N.F.; Donowitz, M.; Estes, M.K. Epithelial WNT Ligands Are Essential Drivers of Intestinal Stem Cell Activation. *Cell Rep.* **2018**, *22*, 1003–1015. [CrossRef] [PubMed]
15. Wu, H.; Xie, S.; Miao, J.; Li, Y.; Wang, Z.; Wang, M.; Yu, Q. Lactobacillus reuteri maintains intestinal epithelial regeneration and repairs damaged intestinal mucosa. *Gut Microbes* **2020**, *11*, 997–1014. [CrossRef] [PubMed]
16. Yao, C.; Gou, X.; Tian, C.; Zhou, L.; Hao, R.; Wan, L.; Wang, Z.; Li, M.; Tong, X. Key regulators of intestinal stem cells: Diet, microbiota, and microbial metabolites. *J. Genet. Genom.* **2023**, *50*, 735–746. [CrossRef]
17. Xie, J.; Li, L.F.; Dai, T.Y.; Qi, X.; Wang, Y.; Zheng, T.Z.; Gao, X.Y.; Zhang, Y.J.; Ai, Y.; Ma, L.; et al. Short-Chain Fatty Acids Produced by Ruminococcaceae Mediate α-Linolenic Acid Promote Intestinal Stem Cells Proliferation. *Mol. Nutr. Food Res.* **2022**, *66*, e2100408. [CrossRef] [PubMed]
18. Duan, C.; Wu, J.; Wang, Z.; Tan, C.; Hou, L.; Qian, W.; Han, C.; Hou, X. Fucose promotes intestinal stem cell-mediated intestinal epithelial development through promoting Akkermansia-related propanoate metabolism. *Gut Microbes* **2023**, *15*, 2233149. [CrossRef] [PubMed]
19. Liu, S.; Du, M.; Tu, Y.; You, W.; Chen, W.; Liu, G.; Li, J.; Wang, Y.; Lu, Z.; Wang, T.; et al. Fermented mixed feed alters growth performance, carcass traits, meat quality and muscle fatty acid and amino acid profiles in finishing pigs. *Anim. Nutr.* **2023**, *12*, 87–95. [CrossRef] [PubMed]
20. Kerr, B.J.; Weber, T.E.; Ziemer, C.J. Dietary marker effects on fecal microbial ecology, fecal VFA, nutrient digestibility coefficients, and growth performance in finishing pigs. *J. Anim. Sci.* **2015**, *93*, 2183–2190. [CrossRef]
21. *GB 5009.246-2016*; Determination of Titanium Dioxide in Food. National Health and Family Planning Commission of the People's Republic of China: Beijing, China, 2016.
22. *GB/T 6435-2014*; Determination of Moisturein Feedstuffs. Standardization Administration of China: Beijing, China, 2014.
23. *GB/T 6432-2018*; Determination of Crude Protein in Feed—Kjeldahl Method for Nitrogen Determination. Standardization Administration of China: Beijing, China, 2018.
24. *GB/T 6433-2006*; Determination of Ether Extract in Feed. Standardization Administration of China: Beijing, China, 2006.
25. *GB/T 6434-2022*; Determination of Crude Fiber Content in Feed. Standardization Administration of China: Beijing, China, 2022.
26. Zhou, J.Y.; Xie, W.W.; Hu, T.C.; Wang, X.F.; Yan, H.C.; Wang, X.Q. Mulberry Leaf-Derived Morin Activates β-Catenin by Binding to Frizzled7 to Promote Intestinal Stem Cell Expansion upon Heat-Stable Enterotoxin b Injury. *J. Agric. Food Chem.* **2024**, *72*, 10366–10375. [CrossRef]

27. Li, X.G.; Zhu, M.; Chen, M.X.; Fan, H.B.; Fu, H.L.; Zhou, J.Y.; Zhai, Z.Y.; Gao, C.Q.; Yan, H.C.; Wang, X.Q. Acute exposure to deoxynivalenol inhibits porcine enteroid activity via suppression of the Wnt/β-catenin pathway. *Toxicol. Lett.* **2019**, *305*, 19–31. [CrossRef] [PubMed]
28. Makkar, H.P.S. Review: Feed demand landscape and implications of food-not feed strategy for food security and climate change. *Animal* **2018**, *12*, 1744–1754. [CrossRef] [PubMed]
29. Chen, J.; Cai, Y.; Wang, Z.; Xu, Z.; Zhuang, W.; Liu, D.; Lv, Y.; Wang, S.; Xu, J.; Ying, H. Solid-state fermentation of corn straw using synthetic microbiome to produce fermented feed: The feed quality and conversion mechanism. *Sci. Total Environ.* **2024**, *920*, 171034. [CrossRef] [PubMed]
30. Shuai, C.; Chen, D.; Yu, B.; Luo, Y.; Zheng, P.; Huang, Z.; Yu, J.; Mao, X.; Yan, H.; He, J. Effect of fermented rapeseed meal on growth performance, nutrient digestibility, and intestinal health in growing pigs. *Anim. Nutr.* **2023**, *15*, 420–429. [CrossRef] [PubMed]
31. Wang, C.; Shi, C.; Su, W.; Jin, M.; Xu, B.; Hao, L.; Zhang, Y.; Lu, Z.; Wang, F.; Wang, Y.; et al. Dynamics of the Physicochemical Characteristics, Microbiota, and Metabolic Functions of Soybean Meal and Corn Mixed Substrates during Two-Stage Solid-State Fermentation. *mSystems* **2020**, *5*, e00501-19. [CrossRef] [PubMed]
32. Gasnier, M.; Lim, H.Y.G.; Barker, N. Role of Wnt signaling in the maintenance and regeneration of the intestinal epithelium. *Curr. Top. Dev. Biol.* **2023**, *153*, 281–326. [CrossRef] [PubMed]
33. Xu, B.; Zhu, L.; Fu, J.; Li, Z.; Wang, Y.; Jin, M. Overall assessment of fermented feed for pigs: A series of meta-analyses. *J. Anim. Sci.* **2019**, *97*, 4810–4821. [CrossRef] [PubMed]
34. Canibe, N.; Jensen, B.B. Fermented liquid feed-Microbial and nutritional aspects and impact on enteric diseases in pigs. *Anim. Feed. Sci. Technol.* **2012**, *173*, 17–40. [CrossRef]
35. Shi, C.; Zhang, Y.; Lu, Z.; Wang, Y. Solid-state fermentation of corn-soybean meal mixed feed with Bacillus subtilis and Enterococcus faecium for degrading antinutritional factors and enhancing nutritional value. *J. Anim. Sci. Biotechnol.* **2017**, *8*, 50. [CrossRef] [PubMed]
36. Sugiharto, S.; Ranjitkar, S. Recent advances in fermented feeds towards improved broiler chicken performance, gastrointestinal tract microecology and immune responses: A review. *Anim. Nutr.* **2019**, *5*, 1–10. [CrossRef] [PubMed]
37. Śliżewska, K.; Markowiak-Kopeć, P.; Sip, A.; Lipiński, K.; Mazur-Kuśnirek, M. The Effect of Using New Synbiotics on the Turkey Performance, the Intestinal Microbiota and the Fecal Enzymes Activity in Turkeys Fed Ochratoxin A Contaminated Feed. *Toxins* **2020**, *12*, 578. [CrossRef]
38. Yu, D.Y.; Oh, S.H.; Kim, I.S.; Kim, G.I.; Kim, J.A.; Moon, Y.S.; Jang, J.C.; Lee, S.S.; Jung, J.H.; Park, J.; et al. Intestinal microbial composition changes induced by Lactobacillus plantarum GBL 16, 17 fermented feed and intestinal immune homeostasis regulation in pigs. *J. Anim. Sci. Technol.* **2022**, *64*, 1184–1198. [CrossRef]
39. Wang, K.; Chen, G.; Cao, G.; Xu, Y.; Wang, Y.; Yang, C. Effects of Clostridium butyricum and Enterococcus faecalis on growth performance, intestinal structure, and inflammation in lipopolysaccharide-challenged weaned piglets. *J. Anim. Sci.* **2019**, *97*, 4140–4151. [CrossRef]
40. Wang, Q.; Fang, Z.; Li, L.; Wang, H.; Zhu, J.; Zhang, P.; Lee, Y.K.; Zhao, J.; Zhang, H.; Lu, W.; et al. Lactobacillus mucosae exerted different antiviral effects on respiratory syncytial virus infection in mice. *Front. Microbiol.* **2022**, *13*, 1001313. [CrossRef]
41. Barker, N.; van Es, J.H.; Kuipers, J.; Kujala, P.; van den Born, M.; Cozijnsen, M.; Haegebarth, A.; Korving, J.; Begthel, H.; Peters, P.J.; et al. Identification of stem cells in small intestine and colon by marker gene Lgr5. *Nature* **2007**, *449*, 1003–1007. [CrossRef] [PubMed]
42. Sato, T.; Vries, R.G.; Snippert, H.J.; van de Wetering, M.; Barker, N.; Stange, D.E.; van Es, J.H.; Abo, A.; Kujala, P.; Peters, P.J.; et al. Single Lgr5 stem cells build crypt-villus structures in vitro without a mesenchymal niche. *Nature* **2009**, *459*, 262–265. [CrossRef] [PubMed]
43. Basak, O.; Beumer, J.; Wiebrands, K.; Seno, H.; van Oudenaarden, A.; Clevers, H. Induced Quiescence of Lgr5+ Stem Cells in Intestinal Organoids Enables Differentiation of Hormone-Producing Enteroendocrine Cells. *Cell Stem Cell* **2017**, *20*, 177–190.e174. [CrossRef] [PubMed]
44. Leung, C.; Tan, S.H.; Barker, N. Recent Advances in Lgr5(+) Stem Cell Research. *Trends Cell Biol.* **2018**, *28*, 380–391. [CrossRef] [PubMed]
45. Humer, E.; Schedle, K. Fermentation of food and feed: A technology for efficient utilization of macro and trace elements in monogastrics. *J. Trace Elem. Med. Biol.* **2016**, *37*, 69–77. [CrossRef]
46. Huang, C.; Tse, T.J.; Purdy, S.K.; Chicilo, F.; Shen, J.; Meda, V.; Reaney, M.J.T. Depletion of cyanogenic glycosides in whole flaxseed via Lactobacillaceae fermentation. *Food Chem.* **2023**, *403*, 134441. [CrossRef] [PubMed]
47. Jha, R.; Berrocoso, J.D. Review: Dietary fiber utilization and its effects on physiological functions and gut health of swine. *Animal* **2015**, *9*, 1441–1452. [CrossRef]

48. Nyström, E.E.L.; Martinez-Abad, B.; Arike, L.; Birchenough, G.M.H.; Nonnecke, E.B.; Castillo, P.A.; Svensson, F.; Bevins, C.L.; Hansson, G.C.; Johansson, M.E.V. An intercrypt subpopulation of goblet cells is essential for colonic mucus barrier function. *Science* **2021**, *372*, eabb1590. [CrossRef] [PubMed]
49. Liang, L.; Liu, L.; Zhou, W.; Yang, C.; Mai, G.; Li, H.; Chen, Y. Gut microbiota-derived butyrate regulates gut mucus barrier repair by activating the macrophage/WNT/ERK signaling pathway. *Clin. Sci.* **2022**, *136*, 291–307. [CrossRef]
50. Hageman, J.H.; Heinz, M.C.; Kretzschmar, K.; van der Vaart, J.; Clevers, H.; Snippert, H.J.G. Intestinal Regeneration: Regulation by the Microenvironment. *Dev. Cell* **2020**, *54*, 435–446. [CrossRef] [PubMed]
51. Yin, J.; Ding, F.; Cheng, Z.; Ge, X.; Li, Y.; Zeng, A.; Zhang, J.; Yan, W.; Shi, Z.; Qian, X.; et al. METTL3-mediated m6A modification of LINC00839 maintains glioma stem cells and radiation resistance by activating Wnt/β-catenin signaling. *Cell Death Dis.* **2023**, *14*, 417. [CrossRef] [PubMed]
52. Fernandez, A.; Huggins, I.J.; Perna, L.; Brafman, D.; Lu, D.; Yao, S.; Gaasterland, T.; Carson, D.A.; Willert, K. The WNT receptor FZD7 is required for maintenance of the pluripotent state in human embryonic stem cells. *Proc. Natl. Acad. Sci. USA* **2014**, *111*, 1409–1414. [CrossRef]
53. Larasati, Y.; Boudou, C.; Koval, A.; Katanaev, V.L. Unlocking the Wnt pathway: Therapeutic potential of selective targeting FZD(7) in cancer. *Drug. Discov. Today* **2022**, *27*, 777–792. [CrossRef]
54. Guo, X.; Wan, P.; Shen, W.; Sun, M.; Peng, Z.; Liao, Y.; Huang, Y.; Liu, R. Fusobacterium periodonticum BCT protein targeting glucose metabolism to promote the epithelial-mesenchymal transition of esophageal cancer cells by lactic acid. *J. Transl. Med.* **2024**, *22*, 401. [CrossRef]

fermentation

MDPI

Article

Biotechnological Effects of *Lactobacillus plantarum*, Cellulase, and Xylanase on Nutritional Quality and Microbial Community Structure of Corn Stover Silage

Jianliang Liu [1,2,3], Mingjian Liu [1,2,3], Panjie Sheng [1,2,3], Chaoran Song [1,2,3], Weiqin Ma [1,2,3], Baochao Bai [1,2,3], Jiayu Zhao [1,2,3], Shuai Du [1,2,3], Gentu Ge [1,2,3], Zhijun Wang [1,4] and Yushan Jia [1,2,3,*]

1 College of Grassland Science, Inner Mongolia Agricultural University, Hohhot 010019, China; liujianliang0202@163.com (J.L.); liumj_nm@163.com (M.L.); shengpanjie@163.com (P.S.); a1198494634@163.com (C.S.); maweiqin0108@126.com (W.M.); bbc66688@163.com (B.B.); zhaojiayu@emails.imau.edu.cn (J.Z.); dushuai_nm@sina.com (S.D.); gegentu@163.com (G.G.); zhijunwang321@126.com (Z.W.)

2 Key Laboratory of Forage Cultivation, Processing and High Efficient Utilization, Ministry of Agriculture, Hohhot 010019, China

3 Key Laboratory of Grassland Resources, Ministry of Education, Hohhot 010019, China

4 National Center of Pratacultural Technology Innovation (Under Preparation), Hohhot 010050, China

* Correspondence: jys_nm@sina.com

Abstract: As animal husbandry advances, the demand for premium feed has seen substantial growth, while the availability of natural forage resources remains limited. Corn stover, characterized by its high yield and rich nutritional content, has become a vital source of roughage. The application of silage technology to corn stover enhances its palatability, improves its nutritional value, and exerts positive effects on livestock production performance. This study aims to evaluate the impact of different additives and their proportions on the nutrient composition, fermentation quality, and microbial community structure of corn stover silage. Specifically, *Lactobacillus plantarum*, cellulase preparations (commercial cellulase enzyme preparations), and xylanase preparations (commercial xylanase enzyme preparations) were employed as additives to investigate their synergistic effects and underlying mechanisms during the silage process. Furthermore, dosage gradients were established to determine the optimal dosage range, providing a robust scientific basis for the optimization of additive applications. In this experiment, corn stover was used as the substrate, with the moisture content maintained at 60%. Treatments with *Lactobacillus plantarum*, cellulase, and xylanase were applied, and silage samples were analyzed after 30 and 60 days of fermentation to assess fermentation quality, nutritional quality, and microbial community structure. The findings revealed that the nutritional quality of corn stover silage improved progressively with fermentation time. Compared to the control group, the addition of *Lactobacillus plantarum*, cellulase, and xylanase significantly increased the abundance of lactic acid bacteria, reduced the pH value, and effectively suppressed the proliferation of spoilage microorganisms. Among the treatments, xylanase demonstrated the most pronounced effects, substantially increasing lactic acid and soluble carbohydrate content while reducing levels of neutral detergent fiber (NDF) and acid detergent fiber (ADF). Notably, the incorporation of 20 U/g xylanase into the silage process facilitated the breakdown of xylan in corn stover into soluble carbohydrates, thereby providing essential substrates for lactic acid bacteria and other beneficial microorganisms. This, in turn, inhibited the growth of harmful microorganisms, ultimately improving the nutritional quality, fermentation quality, and microbial community structure of the silage. These findings provide a theoretical framework and practical guidance for optimizing the production of corn stover silage, promoting efficient resource utilization, and supporting the sustainable development of animal husbandry.

Keywords: cellulase; xylanase; silage execution; corn stalks; straw feedstuff

1. Introduction

With the rapid development of China's animal husbandry industry, the demand for high-quality feed has been steadily increasing. However, the scarcity of natural forage resources and the limited biological yield present significant challenges in meeting the growing demand for roughage in the country's animal husbandry sector [1,2]. In recent years, rising international prices of corn, soybeans, and other food crops have led to a gradual shift in feed composition, from being entirely food-based to a blend of both food and forage [3]. Corn stover, as an essential source of roughage, has emerged as a key solution to the shortage of feed resources due to its high yield, rich nutritional content, and widespread availability. However, directly feeding corn stover to livestock and poultry without prior treatment often limits its broader application due to its high hardness, poor palatability, and low digestibility [4]. From an academic perspective on silage, processing corn stover through silage fermentation not only enhances its palatability but also effectively preserves its nutritional components. This approach significantly improves the feed's nutritional value and contributes to optimizing livestock production performance [5].

Straw silage refers to the process of converting raw straw into a nutrient-rich, easily preserved, and palatable feed through anaerobic fermentation by microorganisms in a sealed environment [6,7]. During ensiling, several factors, including the moisture content of the raw materials, temperature and oxygen conditions in the silage environment, and the use of additives, significantly influence microbial metabolic activity and the final quality of the fermented feed [8]. Appropriate silage additives not only promote the growth and proliferation of lactic acid bacteria, accelerating the anaerobic fermentation process, but also inhibit the growth of harmful microorganisms, thereby improving both the nutritional value and fermentation quality of the silage [9]. The mechanisms of various additives in corn straw silage vary. Some additives enhance feed digestibility by breaking down straw cellulose [10], while others improve silage quality by regulating pH levels in the fermentation environment and suppressing the growth of undesirable microorganisms [11,12].

Previous research on corn stalk silage has predominantly focused on the selection of additive types, but limited attention has been given to the combined effects of different additives and their dosages on the nutritional quality, fermentation characteristics, and microbial community structure of silage. Additionally, there is a lack of reviews on studies that have investigated similar enzymes and microorganisms, such as *Lactobacillus* species, used to modify silage quality. Various species of *Lactobacillus* have been shown to play a vital role in improving silage quality due to their ability to ferment sugars and produce lactic acid, which lowers pH and inhibits the growth of harmful microorganisms [13]. Previous studies have shown that *Lactobacillus plantarum* enhances both the nutritional and fermentation quality of silage by increasing lactic acid production and improving microbial stability [14].

Research on corn stalk silage has predominantly focused on the selection of additive types, while limited attention has been given to the combined effects of different additives and their dosages on the nutritional quality, fermentation characteristics, and microbial community structure of silage. This study employs corn stalks harvested in 2023 as experimental materials to systematically examine the influence of various additives and their dosages on silage quality. Specifically, three additives—*Lactobacillus plantarum*, cellulase, and xylanase—are analyzed across a range of dosages to explore their individual and synergistic effects, as well as the underlying mechanisms driving these interactions. Furthermore, dosage gradients are systematically evaluated to identify the optimal dosage ranges, providing a robust scientific foundation for enhancing the nutritional quality of corn stalk silage. This research aims to offer theoretical insights and practical guidance for the production of high-quality silage, support the advancement and dissemination of silage technologies, address feed resource shortages, and contribute to the sustainable and efficient development of the livestock industry.

2. Materials and Methods

2.1. Test Materials

The corn straw used as silage material was harvested in the autumn from the same experimental field located in Lindong Town, Balinzuo Banner, Chifeng City, Inner Mongolia Autonomous Region, China (43°56′ N, 119°22′ E). The straw was shredded into 2–3 cm pieces and thoroughly mixed to achieve a consistent moisture content of 60%.

Cellulase, mainly including exo-β-glucanase and endo-β-glucanase (CE activity of 10^5 U/g); and xylanase, mainly including β-Xylosidase and Lignase, (XE activity of 10^5 U/g), were purchased from Beijing Xiasheng Biological Technology Development Co., Ltd. (Beijing, China), whereas *Lactobacillus plantarum* (LAB application rate: 1×10^6 cfu/g FM) was obtained from Gansu Punuobeikang Biotechnology Co., Ltd. (Lanzhou, China).

2.2. Experimental Design

Five silage additions and one control group were designed as follows: (1) adding 1×10^5 cfu/g of fresh material (FM) with commercial *Lactobacillus plantarum* fermentation group (LP); (2) adding 1×10^5 cfu/kg FM with cellulase fermentation group (LPCE1); (3) adding 2×10^5 U/kg FM with cellulase fermentation group (LPCE2); (4) adding 1×10^5 U/kg FM with xylanase fermentation group (LPXE1); (5) adding 2×10^5 U/kg FM with xylanase fermentation group (LPXE2); the reference group (CK) received the same amount of distilled water without additives. A total of 1.5 kg of crushed corn straw was weighed, and the additive solution for each treatment was evenly sprayed onto the straw using a sterile sprayer. The corn straw was thoroughly mixed and placed into polyethylene silage bags, with 500 g per bag. The bags were vacuum-sealed, and the number of replications for each treatment was three. Fermentation was carried out at room temperature (20–25 °C), and the bags were opened at 30 and 60 days to assess the nutritional quality, fermentation quality, and microbial community composition.

2.3. Indicators and Methods of Determination

2.3.1. Fermentation Index

A 20 g corn straw sample was homogenized with 180 mL of deionized water at 12 times per second for 2 min, then filtered. The pH value, and lactic acid (LA), acetic acid (AA), propionic acid (PA), butyric acid (BA), and ammonia nitrogen (NH_3-N) contents of the corn straw were determined. The pH value was measured using a pH meter (PHS-3C, INESA Scientific Instrument Co., Ltd., Shanghai, China) [15]. Ammonia nitrogen (NH_3-N) was determined by phenol–sodium hypochlorite colorimetry [16]. High-performance liquid chromatography (HPLC, Waters, Milford, MA, USA, Model: E2695) [17] was used to determine the contents of LA, AA, PA, and BA. The HPLC conditions were as follows: column—Shodex R Spak KC-811S-DVB Gel Column (30 mm × 8 mm); detector—2489 UV–vis detector; mobile phase—3 mol/L perchloric acid; flow rate—1 mL/min; column temperature—50 °C; detection wavelength—210 nm; injection volume—5 μL.

2.3.2. Nutritional Quality

The dry matter content was determined using the drying method [18]. New or silage tests were put in a forced-air broiler at 65 °C for more than 72 h until a steady weight was accomplished. The water-soluble carbohydrate (WSC) content was measured utilizing anthrone colorimetry [19]. The acid detergent fiber (ADF) and neutral detergent fiber (NDF) [20] were measured utilizing the ANKOM A220i programmed fiber analyzer (ANKOM Innovation Corp., Fairport, NY, USA). The crude protein (CP) substance was measured utilizing the Dumas combustion method [21].

2.4. DNA Extraction and PCR Amplification

Total microbial genomic DNA was extracted from new corn stalk tests and silage tests utilizing the E.Z.N.A.® DNA Unit (Omega Bio-tek, Norcross, GA, USA), according to the manufacturer's instructions. The quality and concentration of the DNA were determined by 1.0% agarose gel electrophoresis and a NanoDrop2000 spectrophotometer (Thermo Scientific, United States) and kept at −80 °C further use. The hypervariable region V3–V4 of the bacterial 16S rRNA quality was intensified with preliminary sets 338F (5′-ACTCCTACGGGAGGCAGCAG-3′) and 806R (5′-GGACTACHVGGGTWTCTAAT-3′) [22] by a T100 Warm Cycler PCR thermocycler (BIO-RAD, Hercules, CA, USA). The PCR reaction mixture including 4 μL of 5 × Quick Pfu buffer, 2 μL of 2.5 mM dNTPs, 0.8 μL each primer (5 μM), 0.4 μL Quick Pfu polymerase, 10 ng of template DNA, and ddH_2O to a final volume of 20 μL. The PCR enhancement cycling conditions were as follows: initial denaturation at 95 °C for 3 min, followed by 27 cycles of denaturing at 95 °C for 30 s, strengthening at 55 °C for 30 s, and expansion at 72 °C for 45 s, and a single expansion at 72 °C for 10 min, and finishing at 4 °C. The PCR product was extracted from 2% agarose gel and cleaned utilizing the PCR Clean-Up Pack (YuHua, Shanghai, China) according to the manufacturer's instructions, and measured utilizing Qubit 4.0 (Thermo Fisher Logical, Waltham, MA, USA). The cleaned amplicons were pooled in equimolar sums and paired-end sequenced on an Illumina PE300 stage (Illumina, San Diego, CA, USA), conforming to the standard conventions from Majorbio Bio-Pharm Innovation Co. Ltd. (Shanghai, China). The crude sequencing reads were stored in the NCBI Sequence Read Archive (SRA) database (Promotion Number: PRJNA1182493).

2.5. Bacterial Community Sequencing Analysis

After demultiplexing, the resulting sequences were quality-filtered with fastp (0.19.6) [23] and merged with FLASH (v1.2.11) [24]. At that point the high-quality groupings were de-noised utilizing the DADA2 [23] plugin within the Qiime2 [25] (adaptation 2020.2) pipeline with prescribed parameters, which obtains a single-nucleotide determination based on error profiles within samples. DADA2-denoised arrangements are ordinarily called amplicon sequence variants (ASVs). To reduce the impact of the sequencing on the alpha- and beta-diversity degree, the number of groupings from each test was rarefied to 20,000, which still yielded a normal Good's scope of 99.95%. Taxonomic assignment of ASVs was performed utilizing the Naive Bayes consensus scientific classification classifier actualized in Qiime2 and the SILVA 16S rRNA database (v138). The metagenomic work was used PICRUSt2 (Phylogenetic Examination of Communities by Reproduction of Imperceptibly States) [26] based on ASV agent groupings. PICRUSt2 is a computer program containing an arrangement of tools as follows: HMMER was utilized to adjust the ASV agent arrangements with reference groupings. EPA-NG and Gappa were utilized to put the ASV agent arrangements into a reference tree. The castor was utilized to normalize the 16S quality duplicates. MinPath was utilized to predict quality family profiles, and find the quality pathways. Entire analysis process was accord to protocols of PICRUSt2.

A bioinformatic analysis of the silage microorganism was carried out using the Majorbio Cloud platform (https://cloud.majorbio.com (accessed on 1 December 2024)). Based on the ASV information, rarefaction curves and alpha diversity indices including observed ASVs, Chao1 richness, Shannon index, and Good's coverage were calculated with Mothur v1.30.1 [27]. The similarity among the microbial communities in different samples was determined by principal coordinate analysis (PCoA) based on Bray–Curtis dissimilarity using the Vegan v2.5-3 package. The PERMANOVA test was used to assess the percentage of variation explained by the treatment along with its statistical significance using the Vegan v2.5-3 package. The linear discriminant analysis (LDA) effect size (LEfSe) [28] (http://huttenhower.sph.harvard.edu/LEfSe (accessed on 1December 2024)) was performed to identify the significantly abundant taxa (phylum to genera) of bacteria among the different groups (LDA score > 2, $p < 0.05$). A distance-based redundancy analysis (db-RDA) was performed using the Vegan v2.5-3 package to investigate the effect of nutritional quality and fermentation quality on the silage bacterial community structure. The forward selection was based on Monte Carlo permutation tests (permutations = 9999). The values of the x- and y-axes and the length of the corresponding arrows represent the importance of each nutritional quality and fermentation quality in explaining the distribution of taxon across communities. Linear regression analysis was applied to determine the association between major nutritional quality and fermentation quality, as identified by db-RDA analysis and microbial alpha diversity indices.

2.6. Data Analysis

The experimental data were analyzed in Microsoft Office Excel 2019, and two-way analysis of variance was carried out using the SAS 9.2 statistical software (SAS Institute 2007, Cary, NC, USA). The statistical model of SAS is: $Yijk = \mu + Ti + Dj + (T \times D)\ ij + \varepsilon ijk$, where Yijk = reaction variable, μ = overall mean, Ti = impact of different treatments (CK, LP, LPCE1, LPCE2, LPXE1, LPXE2), Dj = impact of silage time (30 d, 60 d), $(T \times D)$ ijk = interaction impact of distinctive treatments and silage time, εijk = random residual. The difference between the results was analyzed by analysis of variance (ANOVA) with Duncan's post hoc test, and $p < 0.05$ was considered statistically significant; Duncan's post hoc test was utilized for different comparisons of each treatment, and $p < 0.05$ showed that the difference was significant.

3. Results and Analysis

3.1. Effects of Additives and Their Proportions on the Nutritional Quality of Corn Stover

3.1.1. Nutritional Quality of Corn Straw Raw Materials

The nutritional quality of the corn straw silage raw materials is shown in Table 1. The dry matter content of corn straw is 37.04% FM, and the contents of ADF, NDF, CP, and WSC are 44.66% DM, 72.81% DM, 3.98% DM, and 3.91% DM, respectively. As shown in Table 1, corn straw has a higher content of ADF and NDF and a lower content of WSC.

Table 1. Nutritional quality of corn stalk silage raw materials.

Item	Dry Matter, DM (%FM)	Acid Detergent Fiber, ADF (%DM)	Neutral Detergent Fiber, NDF (%DM)	Crude Protein, CP (%DM)	Water-Soluble Carbohydrates, WSCs (%DM)
Corn Stalks	37.04 ± 2.32	48.66 ± 1.73	72.81 ± 0.4	8.48 ± 0.47	8.9 ± 0.64

3.1.2. Effects of Different Additive Treatments on the Nutritional Quality of Corn Straw Silage

The dynamic changes in the nutritional quality of corn straw silage are shown in Table 2. No significant differences were observed in the nutritional quality of the corn straw silage between different additive treatments and silage days ($p > 0.05$). However, significant differences were found in the DM and WSC contents of the corn straw among different additive treatments ($p < 0.05$), while silage time significantly affected the CP, ADF, NDF, and WSC contents ($p < 0.05$). After 60 days of ensiling, the DM content of the corn straw in the LP and LPXE2 treatment groups was 39.84% and 39.06%, respectively; these were significantly lower than that in the CK treatment group ($p < 0.05$). The CP content in the CK group was significantly lower than in all other additive treatment groups ($p < 0.05$). After 30 days of ensiling, the ADF contents of LPCE1, LPCE2, LPXE1, and LPXE2 were 44.21% DM, 41.17% DM, 45.02% DM, and 43.84% DM, respectively, which were significantly lower than those in the CK and LP groups ($p < 0.05$). The addition of cellulase and xylanase significantly reduced the ADF content in corn straw silage, and the cellulase concentration had a significant effect on the ADF content ($p < 0.05$). After 30 days of silage, the NDF contents in the LPCE1, LPXE1, and LPXE2 treatment groups were 68.36% DM, 65.84% DM, and 64.82% DM, respectively, significantly lower than those in the CK and LP groups ($p < 0.05$). No significant differences were found between the CK and LP treatment groups ($p > 0.05$). After 60 days of ensiling, the NDF content ranged from 69.57% to 65.21% DM. The NDF contents in the CK and LP treatment groups were significantly higher than in the LPXE1 and LPXE2 treatment groups ($p < 0.05$). A high concentration of xylanase significantly reduced the NDF content after 60 days of ensiling. Compared with the CK and LP groups, the WSC contents in the LPCE1, LPCE2, LPXE1, and LPXE2 groups were significantly higher ($p < 0.05$). High concentrations of cellulase and xylanase significantly enhanced the WSC content, while silage duration had no significant effect on WSC content ($p > 0.05$).

Table 2. Impacts of diverse added substances on the dietary quality of corn straw silage.

	Time	CK	LP	LPCE1	LPXE1	LPCE2	LPXE2	SEM	T	D	T × D
DM%FM	30 d	41.78 ^{Aa}	40.39 ^{Aab}	40.22 ^{Ab}	41.41 ^{Aab}	40.67 ^{Aab}	40.43 ^{Aab}	0.1624	<0.0001	0.0349	0.8293
	60 d	41.37 ^{Aa}	39.84 ^{Abc}	40.10 ^{Aab}	40.82 ^{Aab}	40.16 ^{Aab}	39.06 ^{Bc}				
CP%DM	30 d	2.56 ^{Ac}	3.31 ^{Aa}	3.10 ^{Aab}	3.00 ^{Aab}	2.83 ^{Bbc}	3.42 ^{Aa}	0.0570	0.0116	<0.0001	0.1453
	60 d	2.64 ^{Ab}	3.22 ^{Aa}	3.34 ^{Aa}	3.38 ^{Aa}	3.33 ^{Aa}	3.42 ^{Aa}				
ADF%DM	30 d	48.2 ^{Aa}	47.76 ^{Aa}	44.21 ^{Ab}	45.02 ^{Ab}	41.17 ^{Ac}	43.84 ^{Ab}	0.4372	0.0945	<0.0001	0.4532
	60 d	48.01 ^{Aa}	47.64 ^{Aa}	43.37 ^{Ab}	42.97 ^{Bb}	41.39 ^{Ac}	43.31 ^{Ab}				
NDF%DM	30 d	70.43 ^{Aa}	70.31 ^{Aa}	68.36 ^{Ab}	65.84 ^{Ac}	70.44 ^{Aa}	64.82 ^{Ac}	0.3681	0.0199	<0.0001	0.375
	60 d	69.26 ^{Aab}	68.36 ^{Bab}	67.62 ^{Ab}	65.58 ^{Ac}	69.57 ^{Aa}	65.21 ^{Ac}				
WSC%DM	30 d	3.11 ^{Ae}	4.14 ^{Ade}	4.94 ^{Acd}	5.50 ^{Abc}	6.37 ^{Aab}	7.01 ^{Aa}	0.2534	<0.0001	<0.0001	0.1011
	60 d	2.73 ^{Ac}	1.96 ^{Bc}	4.98 ^{Ab}	5.35 ^{Ab}	5.87 ^{Aab}	7.17 ^{Aa}				

Note: (1) DM, dry matter; WSC, water-soluble carbohydrate; ADF, acid detergent fiber; NDF, neutral detergent fiber; CP, crude protein; FM, fresh material; CK, no additive contro; LP, materials. (2) Different lowercase letters within the same row show significant differences ($p < 0.05$) between different treatments at the same ensiling duration; different capitalized letters within the same column show significant differences ($p < 0.05$) between different ensiling lengths under the same treatment, whereas the same capitalized letters show no significant difference between treatments. T: different additive treatments; D: different ensiling lengths; T × D: interaction between additive treatments and ensiling lengths; SEM: standard error of mean.

3.1.3. Effects of Different Treatments on the Fermentation Quality of Corn Straw Silage

The fermentation quality of corn straw silage after 30 and 60 days of treatment with different additives is shown in Table 3. The interaction between different additives and silage duration significantly affected the pH, NH_3-N, lactic acid (LA), and propionic acid (PA) contents ($p < 0.05$). Silage duration significantly influenced the pH, LA, NH_3-N, and PA ($p < 0.05$), and different additives significantly affected the pH, NH_3-N, LA, acetic acid (AA), and PA of corn straw silage ($p < 0.05$). After 60 days of ensiling with additives, the NH_3-N content was significantly lower than in the CK group ($p < 0.05$). Compared with the CK group, the pH, LA, PA, and AA contents of the corn straw silage with additives were significantly lower after 60 days of ensiling ($p < 0.05$). Additionally, butyric acid (BA) was not detected in any of the treatment groups. No significant differences in NH_3-N, LA, and PA were observed among the LPCE1, LPXE1, LPCE2, and LPXE2 treatment groups ($p > 0.05$). The LA content was highest in the LPCE2 and LPXE2 groups, at 22.24 g/kg FM and 26.74 g/kg FM, respectively. PA was only detected in the CK, LPCE2, and LP groups, and was not detected in the LPCE1, LPXE1, and LPXE2 groups.

Table 3. Effects of different additives on the fermentation quality of corn stover.

	Time	CK	LP	LPCE1	LPXE1	LPCE2	LPXE2	SEM	T	D	T × D
pH	30 d	3.94 ^{Aa}	3.76 ^{Ab}	3.720 ^{Abc}	3.70 ^{Abc}	3.74 ^{Abc}	3.69 ^{Ac}	0.0061	0.4667	0.0016	0.0378
	60 d	3.78 ^{Ba}	3.69 ^{Ab}	3.70 ^{Ab}	3.70 ^{Ab}	3.71 ^{Ab}	3.65 ^{Ac}				
NH_3-N%	30 d	0.7985 ^{Aa}	0.0397 ^{Ab}	0.0275 ^{Ab}	0.0263 ^{Ab}	0.0267 ^{Ab}	0.0213 ^{Ab}	0.0089	<0.0001	<0.0001	<0.0001
	60 d	0.1562 ^{Ba}	0.0989 ^{Aab}	0.045 ^{Abc}	0.0545 ^{Abc}	0.0308 ^{Abc}	0.0200 ^{Ac}				
LA (g/kgFM)	30 d	12.34 ^{Ac}	14.74 ^{Abc}	14.35 ^{Bbc}	14.21 ^{Abc}	16.01 ^{Bb}	19.16 ^{Ba}	0.7495	<0.0001	<0.0001	0.2251
	60 d	14.39 ^{Ab}	15.77 ^{Ab}	17.10 ^{Aa}	18.42 ^{Aa}	22.24 ^{Aa}	26.74 ^{Aa}				
AA (g/kgFM)	30 d	1.14 ^{Ba}	0.76 ^{Ab}	0.63 ^{Abc}	0.56 ^{Abc}	0.61 ^{Abc}	0.53 ^{Ac}	0.0434	<0.0001	0.00014	0.0015
	60 d	1.28 ^{Aa}	0.86 ^{Ab}	0.67 ^{Ac}	0.42 ^{Ae}	0.58 ^{Acd}	0.56 ^{Ad}				
PA (g/kgFM)	30 d	1.07 ^{Ba}	0.84 ^{Bb}	0.00 ^{Ad}	0.00 ^{Ad}	0.58 ^{Bc}	0.00 ^{Ad}	0.1063	<0.0001	<0.0001	<0.0001
	60 d	1.89 ^{Aa}	1.37 ^{Ab}	0.00 ^{Ad}	0.00 ^{Ad}	0.73 ^{Ac}	0.00 ^{Ad}				

Note: (1) CK, no additive contro; LP, materials, NH_3-N: ammonia nitrogen; LA: lactic acid; AA: acetic acid; PA: propionic acid. (2) Different lowercase letters within the same row show significant differences between different treatments at the same silage days ($p < 0.05$). (3) Different capitalized letters within the same column show significant differences between different silage days for the same treatment ($p < 0.05$); the same capitalized letters demonstrate no significant differences between treatments. (4) T: different additive treatments; D: different silage days; T × D: interaction between additive treatment and silage days; SEM: standard error of mean.

3.2. *Effects of Different Additives on Microbial Community of Corn Stover*

3.2.1. Effects of Different Additives on Microbial Community Structure of Corn Stalk Silage

Effects of Different Additives on Microbial Community Diversity of Corn Stalk Silage

Based on the analysis of the nutritional and fermentation quality of corn straw under different additive treatments, this study also conducted microbial research on the CK, LP, LPCE2, and LPXE2 groups. The α-diversity of microbial communities after 30 and 60 days of silage is shown in Figure 1. In this study, the coverage of all silage treatment groups exceeded 99%, ensuring that all microorganisms in the samples were adequately captured. As shown in Figure 1A,B, the addition of cellulase and xylanase during corn straw silage did not significantly affect the microbial diversity indices ($p > 0.05$), and the silage duration had no significant effect on the α-diversity of the microbial communities ($p > 0.05$). After 60 days of silage, there was no significant difference in community richness between the CE2 and XE2 groups ($p > 0.05$); however, the community richness of the CE2 and XE2 groups was significantly higher than that of the CK group ($p < 0.05$). Both the ACE and Chao1 indices showed similar trends. As shown in Figure 1C, after 60 days of silage, the Shannon index of the CK2 group was significantly lower than that of the LP2, CE2, and XE2 groups. Figure 1D shows that the Simpson index of the XE2 group had the lowest diversity, which was significantly different from the CK2 group ($p < 0.05$).

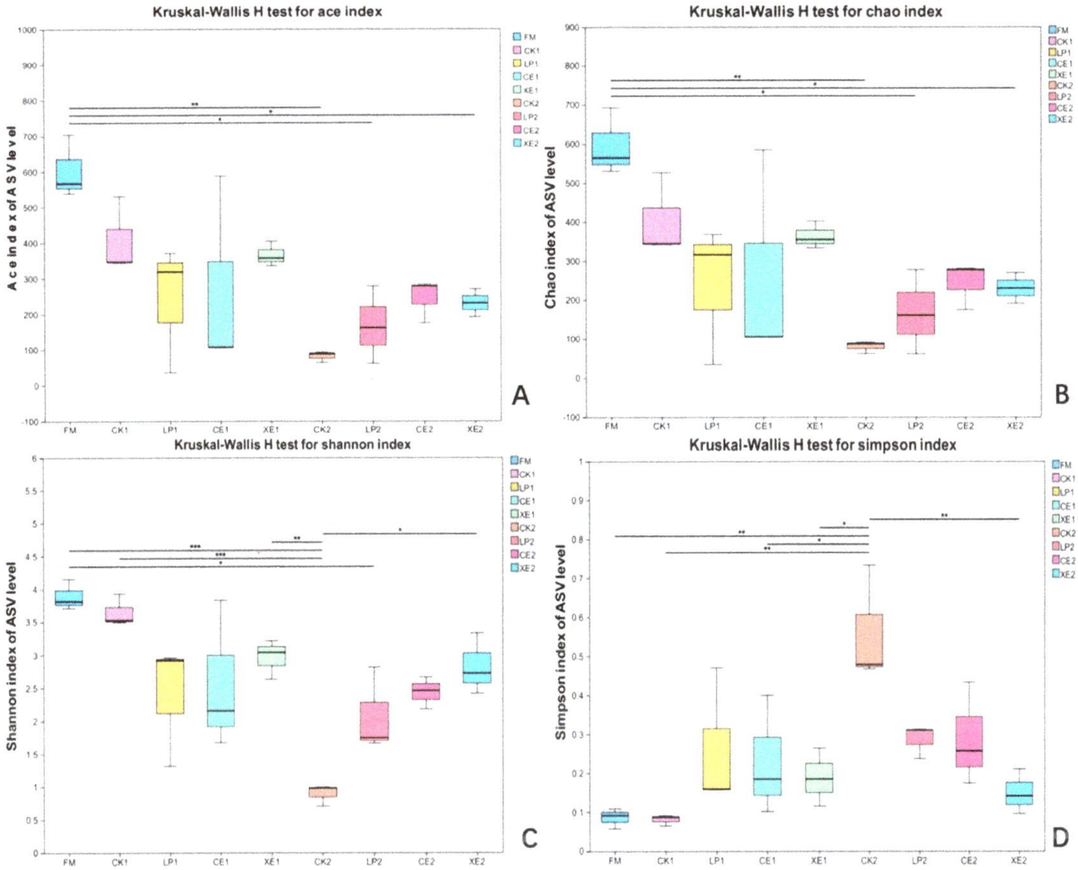

Figure 1. Impacts of different added substances on α-diversity of microorganisms in corn stover silage. The figure shows the significant differences between the selected two groups of samples and marks the two groups with significant differences ($0.01 < p \leq 0.05$ marked as *, $0.001 < p \leq 0.01$ marked as **, $p \leq 0.001$ marked as ***). The x-axis shows the group names, and the y-axis shows the values of each group's pointers. FM: fresh material; CK1: CK aging for 30 days; CK2: CK maturation for 60 days; LP1: LP maturation for 30 days; LP2: LP aging for 60 days; CE1: LPCE2 maturation for 30 days; CE2: LPCE2 maturation for 60 days; XE1: LPXE2 aging for 30 days; XE2: LPXE2 maturation for 60 days. These two numbers are the same. (**A**) Microbial diversity (Ace) of corn straw under different treatment groups, (**B**) Microbial diversity (Chao) of corn straw under different treatment groups, (**C**) Microbial diversity (Shannon) of corn straw under different treatment groups, (**D**) Microbial diversity (Simpson) of corn straw under different treatment groups.

Bata Diversity Analysis

The PCoA examination of corn straw silage at the class level is shown in Figure 2. Noteworthy contrasts were seen within the microbial community structure between new and silage tests. After 30 and 60 days of ensiling, the microbial community structure of the XE group showed no noteworthy contrasts, showing that the microbial community structure within the XE treatment group remained steady all through the whole aging process. In contrast, the microbial communities in the CK, LP, and CE groups showed significant changes, with the CK group exhibiting the most pronounced shifts. After 60 days of silage, a larger degree of dispersion was observed among the three replicates of the LP group, indicating considerable variation in the microbial community structure distribution across different replicates in the LP group after 60 days of fermentation.

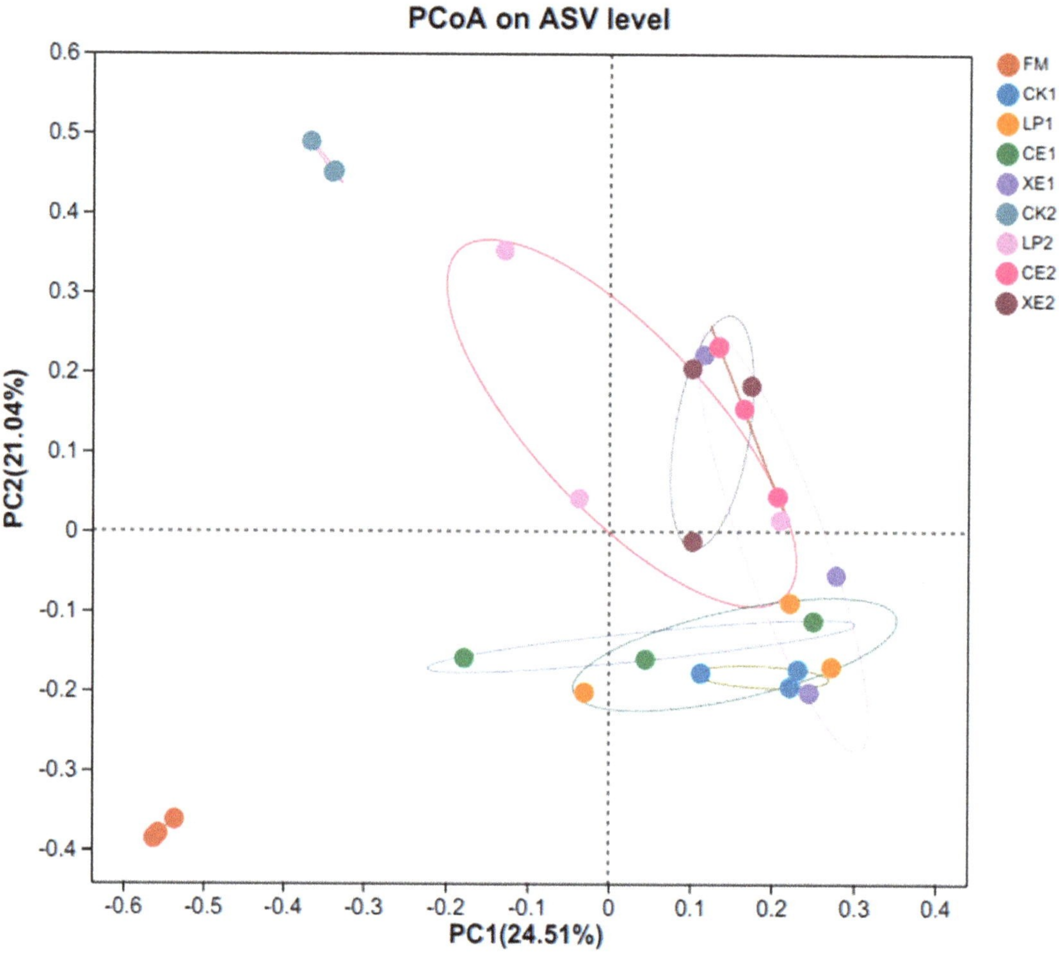

Figure 2. Effects of different additives on β-diversity of corn stalk silage. Note: FM, fresh material; CK1, CK silage fermented for 30 days; CK2, CK silage fermented for 60 days; LP1, LP silage fermented for 30 days; LP2, LP silage fermented for 60 days; CE1, LPCE2 silage fermented for 30 days; CE2, LPCE2 silage fermented for 60 days; XE1, LPXE2 silage fermented for 30 days; XE2, LPXE2 silage fermented for 60 days. he ellipse represents the within-group differences.

Effects of Different Additives on Microbial Community Phylum Level and Genus Level of Corn Straw Silage

Figure 3A,B show the changes within the microbial community structure at the phylum level for different additive-treated groups after corn straw silage. As shown within the figures, the microbial communities in corn straw silage from different aging periods were dominated by the phylum Firmicutes, followed by Proteobacteria, with other phyla showing generally lower bacterial abundance. After 60 days of silage, the relative abundance of Firmicutes had increased within the CK and LP groups compared to 30 days of silage

($p < 0.05$). Within the CE2 group, the relative abundance of Firmicutes was essentially lower than that within the CE1 group ($p < 0.05$), whereas no significant changes were seen within the XE group ($p > 0.05$). Figure 3C,D show the changes within the microbial community structure at the class level, which are completely different to the additive-treated groups after corn straw silage. As shown within the figures, after 60 days of silage, *Lactobacillus* was the foremost class in all treatment groups, except for the CE group. The second most abundant class in these groups was the unclassified_f_*Alcaligenaceae*, and the third was *Paenibacillus*. After 60 days of silage, the relative abundance of *Lactobacillus* within the CK, LP, CE, and XE groups accounted for 97.33%, 76.44%, 50.97%, and 65.63%, respectively. The relative abundance of *Lactobacillus* increased within the CK and XE groups after 60 days of silage ($p < 0.05$), whereas it diminished within the CE group ($p < 0.05$).

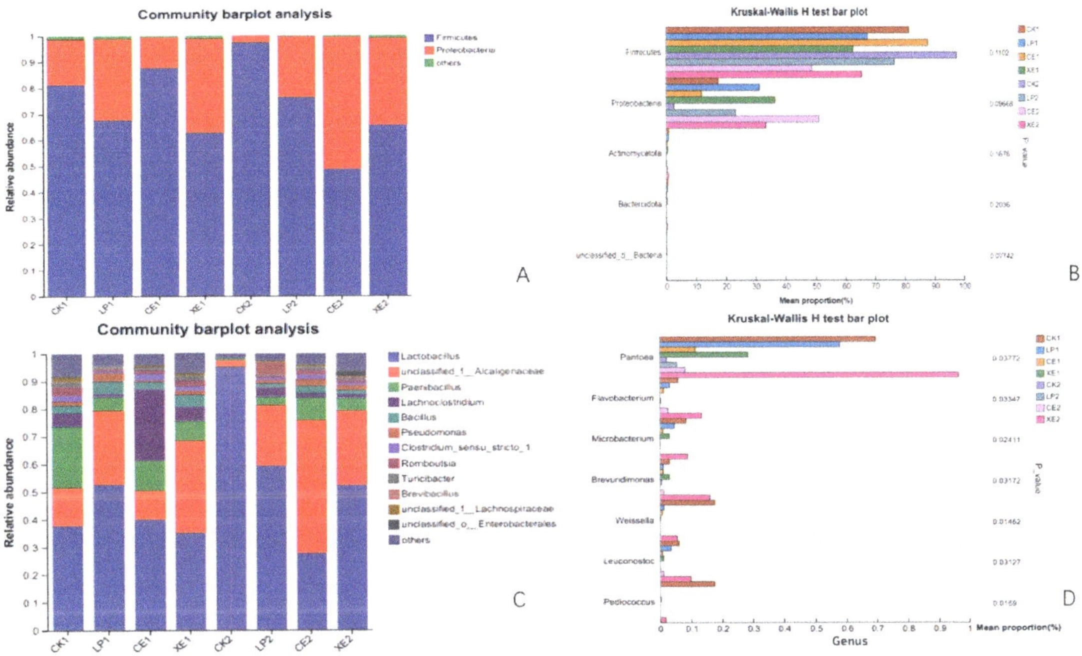

Figure 3. Impacts of different additive treatments on microbial community structure of corn straw silage at phylum and class levels. (**A**) Microbial species composition at phylum level; (**B**) differences in microbial species composition between different treatments at the phylum level; (**C**) microbial species composition at sort level; (**D**) differences in microbial species composition between different treatments at the class level.

3.3. Relationship Investigation of Bacterial Communities with Dietary and Aging Quality of Corn Straw Silage

As shown in Figure 4, the relative abundance of *Lactobacillus* was essentially positively related with NDF, ADF, AA, Dad, and DM content ($p < 0.05$), and essentially negatively connected with NH_3-N, WSC, CP, and LA content ($p < 0.05$). The relative abundances of unclassified_f_*Alcaligenaceae*, *Lachnoclostridium*, and *Paenibacillus* were significantly positively correlated with NH_3-N, WSC, CP, and LA contents, and significantly negatively correlated with NDF, ADF, AA, PA, and DM contents ($p < 0.05$). Additionally, the RDA results indicated that the contents of ADF, NDF, WSC, and CP had a more significant impact on the microbial composition under different treatments.

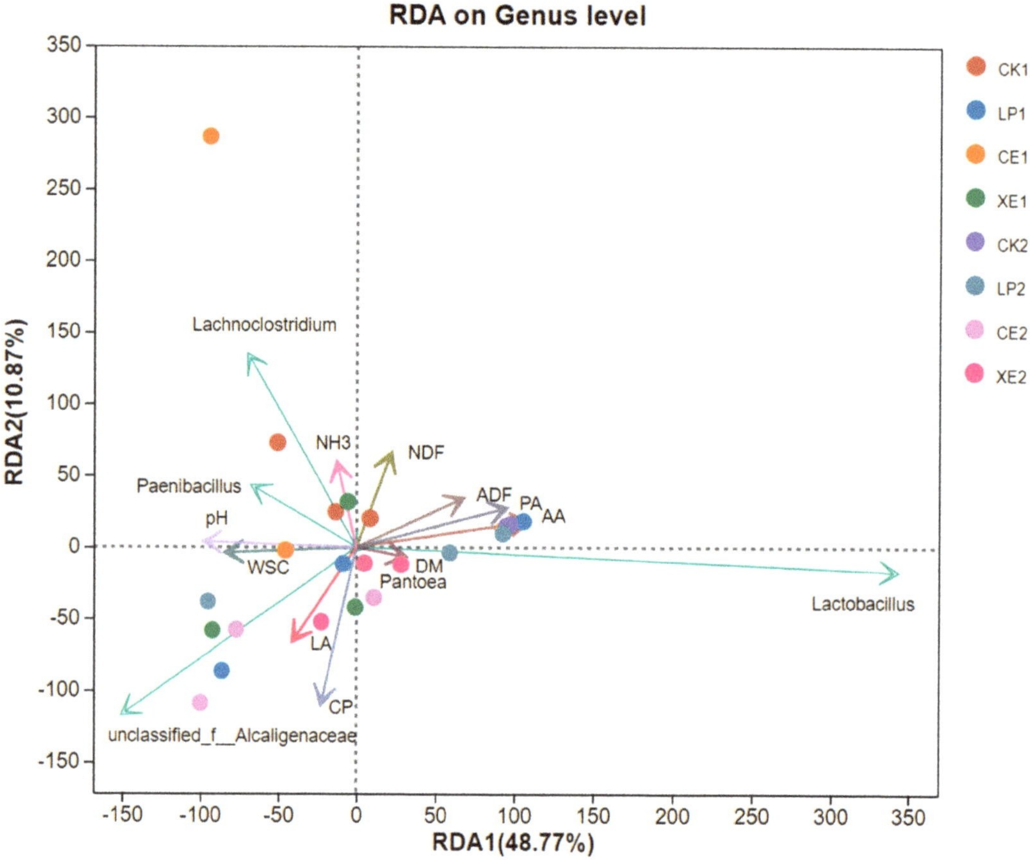

Figure 4. Correlation analysis (RDA) of bacterial communities with nutritional and fermentation quality of corn straw silage. A redundancy analysis (RDA) ordination plot of bacteria and the explanatory variables (such as pH, NH_3-N, LA, AA, DM, NDF, ADF, ADL, CP, WSC) is shown. The variables are represented as arrows. pH: pH value; LA: lactic acid; AA: acetic acid; NH_3-N: ammonia nitrogen; DM: dry matter; NDF: neutral detergent fiber; ADF: acid detergent fiber; CP: crude protein; WSC: water-soluble carbohydrate.

3.4. Prediction of Metabolic Pathways at Three Levels for Corn Straw Silage

The 16S rRNA gene functional prediction analysis for fresh corn straw samples and corn straw samples after 60 days of silage fermentation is shown in Figure 5. At the primary metabolic pathway level (Figure 5A), the overall trend across all treatments remained consistent, with significant enrichment in metabolic processes. Significant differences in metabolism were observed among the treatments. The LP, CE, and XE treatments significantly promoted metabolic pathways, while the CK treatment inhibited these pathways. After silage, all treatments, except for the CK group, showed enhanced metabolic activity. At the secondary metabolic pathway level (Figure 5B), the main metabolic pathways included global metabolism, carbohydrate metabolism, amino acid metabolism, membrane transport, energy metabolism, and metabolism of cofactors and vitamins. After ensiling, the secondary metabolic pathways for the different additive treatments were consistent with those observed at the primary metabolic pathway level.

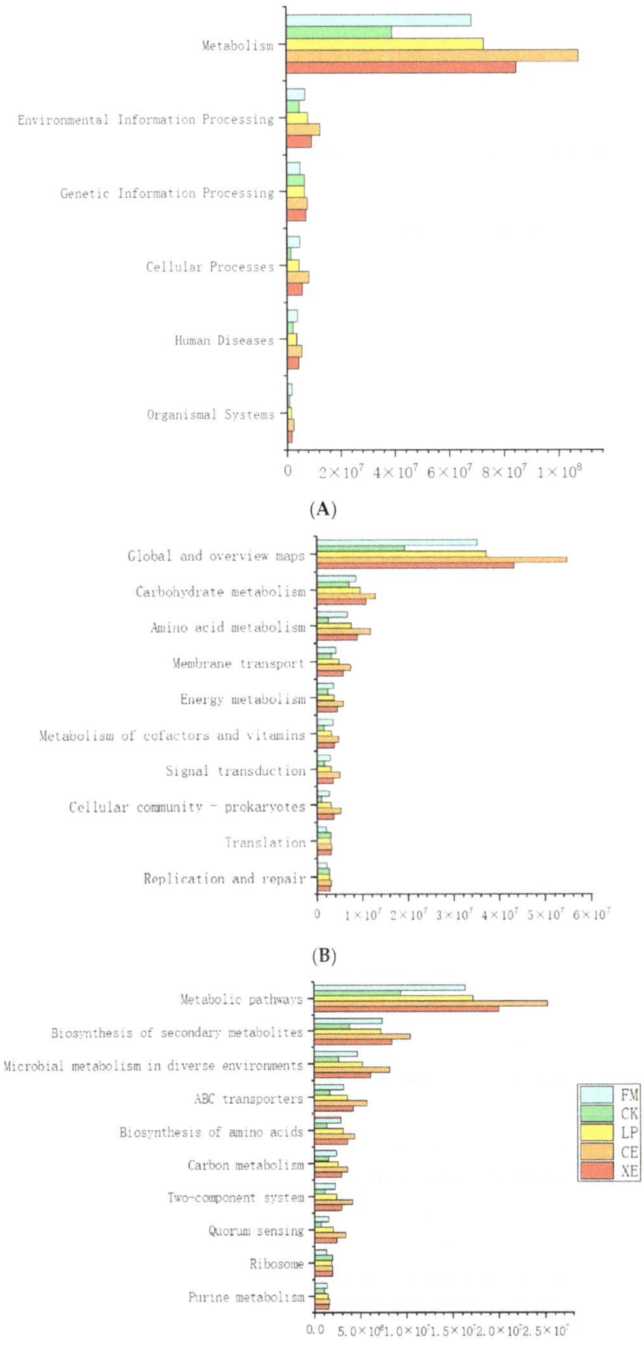

Figure 5. Forecast of 16S rRNA quality work of broom sorghum straw crude materials and silage tests after 60 days of aging. (**A**) Essential pathway level, (**B**) auxiliary pathway level, (**C**) tertiary pathway level; FM: fresh material; CK: control group; LP: *Lactobacillus plantarum* treatment; CE: cellulase treatment; XE: xylanase treatment.

The top 10 tertiary metabolic pathways are shown in Figure 5C. The relative abundance of these pathways was significantly higher than that of other metabolic pathways. The CE-, LP-, and XE-treatment groups promoted most of the metabolic pathways, including metabolic pathways, biosynthesis of secondary metabolites, and microbial metabolism in diverse environments. In contrast, the CK group inhibited metabolic pathways and microbial metabolism in diverse environments, but promoted the ribosome pathway.

4. Discussion

4.1. The Effects of Different Additives on the Nutritional Quality of Corn Straw Silage

During the silage process, the nutritional quality of the silage material undergoes changes due to a series of biological activities, influenced by both the raw material itself and the microorganisms involved [29,30]. In this study, after 60 days of ensiling, the crude protein (CP) content in the additive-treated groups was significantly higher than that in the control (CK) group. This finding aligns with the results of Xu et al. [31], who examined the effect of cellulase on whole-plant corn silage. The improvement in CP content may be attributed to cellulase accelerating the breakdown of cellulose, which provides additional fermentation substrates for lactic acid bacteria. These bacteria, in turn, produce more lactic acid (LA), lowering the pH, inhibiting proteinases from hydrolyzing proteins, and suppressing the growth of harmful microorganisms that also degrade proteins [32]. As a result, the fermentation quality of the corn straw silage is enhanced. Muhammad et al. [33] reported that the addition of cellulase and xylanase to barley straw silage significantly reduced the acid detergent fiber (ADF) content. In the present study, after 60 days of silage, the ADF content in samples treated with both low and high concentrations of cellulase was significantly lower than in the CK group. This result is consistent with previous studies and may be attributed to cellulase degrading the cell walls of corn straw, leading to the release of more soluble sugars. In terms of nutritional quality, the group treated with xylanase showed the best results.

4.2. The Effects of Different Additives on the Fermentation Quality of Corn Straw Silage

The pH value, NH_3-N, and lactic acid (LA) content are key fermentation indicators used to assess silage quality. Typically, a lower pH value correlates with a higher LA content, signifying better fermentation quality [34]. In this study, after 60 days of corn straw silage, the groups treated with cellulase and xylanase effectively reduced the pH value and significantly increased the LA content. Among these treatments, the xylanase treatment proved more effective than the cellulase and *Lactobacillus treatments*. Zhao et al. [35] observed that adding cellulase (100 U/g) to a mixture of corn straw and soybean residue resulted in a significant increase in both lactic acid and acetic acid content after 56 days of silage. Similarly, Li et al. [36] reported that after 60 days of silage at room temperature, corn silage treated with cellulase and xylanase showed significantly higher LA content and lower pH values than the control group. This improvement is likely due to the ability of cellulase to degrade lignin and cellulose, releasing more soluble sugars for lactic acid bacteria fermentation. This, in turn, promotes the activity of homofermentative lactic acid bacteria, inhibits the growth of harmful microorganisms [37,38], lowers the pH, and increases lactic acid production [39,40]. The NH_3-N content reflects the extent to which proteins and amino acids in the silage are broken down by microorganisms. A lower NH_3-N ratio indicates reduced nitrogen loss, which results in higher-quality silage [41]. The addition of cellulase and xylanase in the silage process can enhance protein content in the feed [42] and minimize protein hydrolysis into ammonia-containing compounds by plant proteinases and microorganisms [43]. He et al. [44] reported that adding cellulase

(50–100 U/g) to mulberry leaf silage significantly reduced NH_3-N content. Similarly, in this study, after 60 days of corn straw silage, the NH_3-N content in the additive-treated groups was significantly lower than in the CK group, indicating that cellulase and xylanase helped reduce protein loss during the silage process. This reduction is likely due to the breakdown of cellulose by cellulase and xylanase, which provides more substrate for lactic acid bacteria fermentation, while inhibiting harmful microorganisms and proteinases that hydrolyze proteins [45]. Therefore, this study highlights the significant advantages of using xylanase as an additive to enhance the nutritional quality of corn stover silage.

4.3. The Effects of Different Additives on the Microbial Community of Corn Straw Silage

Silage fermentation is a process in which lactic acid bacteria, naturally present on plant materials, ferment the substrate to produce lactic acid, thereby lowering the pH. During this process, lactic acid bacteria rapidly proliferate by utilizing the nutrients in the raw material, becoming the dominant microbial group, while suppressing the growth of harmful microorganisms such as clostridia and molds. Therefore, studying the microbial diversity and community structure in silage is crucial for understanding and improving the fermentation quality of silage feed. Zhao et al. [35] found that adding cellulase to a mixture of corn straw and soybean residue silage resulted in an increase in both the Shannon and Simpson indices in the silage samples. The results of this study indicate that after 60 days of silage, the Ace, Chao1, and Shannon indices of the additive-treated corn straw silage were significantly lower than those of the fresh material (FM) group. These findings align with those of Zhao. This may be because the additives provide substrates that promote lactic acid bacteria fermentation, which in turn enhances lactic acid production, lowering the pH of the silage. As a result, the complex microbial community in the silage is gradually suppressed by the dominance of lactic acid bacteria fermentation.

Additionally, the application of xylanase during silage fermentation leads to the production of xylooligosaccharides (XOSs). XOSs act as a prebiotic, having a positive impact on the growth of beneficial microorganisms such as lactic acid bacteria. This can further promote the fermentation process and improve the overall fermentation quality of the silage. XOSs not only directly stimulate beneficial microbial growth but also indirectly enhance silage quality by maintaining a favorable microbial environment, contributing to more efficient fermentation and better preservation of nutrients [46].

The fermentation quality of silage feed is closely linked to the microbial community composition within the silage. In this study, after 30 days of silage fermentation, the microbial community in corn straw silage was primarily composed of Firmicutes and Proteobacteria. After 60 days of fermentation, Proteobacteria were gradually replaced by Firmicutes in the microbial community. This shift is likely due to the anaerobic and acidic environment, which favors the dominance of *Lactobacillus* (a genus within Firmicutes) during fermentation, inhibiting the growth of Proteobacteria [47]. At the genus level, unclassified_f_*Alcaligenaceae* was found in all treatment groups. This may be attributed to the addition of cellulase and xylanase, which break down cellulose and lignin, providing soluble sugars as fermentation substrates for microbial communities attached to the surface of corn straw. Consequently, unclassified_f_*Alcaligenaceae* exhibited a higher relative abundance after silage fermentation. Relevant studies suggest that unclassified_f_*Alcaligenaceae* can degrade lignin-derived compounds in the anaerobic fermentation system of silage. This capability may contribute to the breakdown of lignin during the silage process, thereby promoting overall fermentation and improving the quality of the silage feed [48].

4.4. Prediction of Microbial Gene Functions in Silage

Predicting the functional changes in bacterial communities is crucial for evaluating their impact on silage quality changes. In this study, the relative abundance of KEGG pathways in the bacterial community was analyzed using the PICRUSt2 method. The results revealed that, at the primary pathway level, metabolism was the most prominent category, indicating that microbial metabolism drives the entire fermentation process during silage production. Microorganisms utilize fermentation substrates from corn straw to convert available compounds into new metabolites. At the secondary pathway level, pathways such as global metabolism, carbohydrate metabolism, amino acid metabolism, membrane transport, energy metabolism, and metabolism of cofactors and vitamins showed high relative abundance. Previous studies have demonstrated that these metabolic pathways are closely related to the fermentation quality of silage feed [32]. To explore the metabolic processes further, we delved into the third-level metabolic pathways. Silage fermentation is driven by lactic acid bacteria, which utilize soluble carbohydrates from the raw material to produce lactic acid. As fermentation progresses, lactic acid bacteria generate various secondary metabolites through a range of metabolic pathways. Additionally, we observed significant differences in the metabolic pathways across different treatment groups, particularly in pathways such as metabolic pathways, biosynthesis of secondary metabolites, and microbial metabolism in diverse environments. These variations underscore the impact of different additives on the metabolic processes that shape the fermentation quality and microbial dynamics of the silage.

5. Conclusions

This study assessed the impacts of changing the amounts of *Lactobacillus plantarum*, cellulase, and xylanase as added substances, together with silage duration, on the nutritional quality, aging quality, and bacterial community composition of corn stover silage. The results demonstrated that the expansion of *Lactobacillus plantarum*, cellulase, and xylanase essentially improved the tactile quality of corn stover silage, balanced the microbial community during maturation, and upgraded the dietary profile of the silage. Particularly, the incorporation of *Lactobacillus plantarum* and xylanase increased the abundance of *Lactobacillus*, brought down the silage pH, and viably stifled the development of decay microorganisms. In comparison to cellulase, xylanase was more effective in increasing lactic acid (LA) and water-soluble carbohydrates (WSCs) while lessening pH, NH_3-N, and propionic acid (Dad) levels. Based on these discoveries, it can be concluded that the addition of 20 U/g xylanase to corn stover silage essentially improves silage quality, maturation quality, and microbial community flow. Hence, the use of xylanase is suggested for optimizing the corn stover silage quality.

Author Contributions: J.L. was responsible for the technique, software advancement, approval, information administration, and planning of the first draft; M.L. contributed to strategy, program, approval, formal examination, and drafting the introductory composition; B.B. conducted the examination; W.M. provided assets; P.S., C.S. and J.Z. revised and altered the original draft; Z.W. also contributed to checking and editing the manuscript; S.D. took part in the survey and editing; G.G. dealt with visualizations; and Y.J. supervised conceptualization, venture administration, and subsidizing procurement. All authors have read and agreed to the published version of the manuscript.

Funding: The National Key R&D Program of China (2022YFD1601203).

Institutional Review Board Statement: This study was conducted in strict understanding with moral rules. We regarded and ensured the rights and protection of the members and guaranteed the privacy of their individual data.

Informed Consent Statement: Not applicable.

Data Availability Statement: The datasets created in the current study are accessible within the [NCBI] store, [https://www.ncbi.nlm.nih.gov/ (accessed on 1 December 2024), PRJNA1182493].

Conflicts of Interest: The authors announce that they have no known competing money-related interface or individual connections that might have showed up to impact the work detailed in this paper.

References

1. Li, Y.; Hua, D.; Xu, H.; Jin, F.; Mu, H.; Zhao, Y.; Fang, X. Acidogenic and methanogenic properties of corn straw silage: Regulation and microbial analysis of two-phase anaerobic digestion. *Bioresour. Technol.* **2020**, *307*, 123180. [CrossRef] [PubMed]
2. Xie, Y.; Xu, S.; Li, W.; Wang, M.; Wu, Z.; Bao, J.; Jia, T.; Yu, Z. Effects of the Application of *Lactobacillus plantarum* Inoculant and Potassium Sorbate on the Fermentation Quality, In Vitro Digestibility and Aerobic Stability of Total Mixed Ration Silage Based on Alfalfa Silage. *Animals* **2020**, *10*, 2229. [CrossRef]
3. Kuglarz, M.; Alvarado-Morales, M.; Dabkowska, K.; Angelidaki, I. Integrated production of cellulosic bioethanol and succinic acid from rapeseed straw after dilute-acid pretreatment. *Bioresour. Technol.* **2018**, *265*, 191–199. [CrossRef]
4. Huang, R.; Cai, B.; Chen, Y.; Zheng, X.; Yang, J.; Ma, C.; Wang, X.; Zhang, F. Bacterial community structure and metabolites after ensiling paper mulberry mixed with corn or wheat straw. *Front. Sustain. Food Syst.* **2024**, *8*, 1356705. [CrossRef]
5. Zhang, Q.; Zhao, M.; Wang, X.; Yu, Z.; Na, R. Ensiling alfalfa with whole crop corn improves the silage quality and in vitro digestibility of the silage mixtures. *Grassl. Sci.* **2017**, *63*, 211–217. [CrossRef]
6. Li Ying, H.; Borjigin, N.; Yu, Z. Effect of inoculants and fibrolytic enzymes on the fermentation characteristics, in vitro digestibility and aflatoxins accumulation of whole-crop corn silage. *Grassl. Sci.* **2017**, *63*, 69–78. [CrossRef]
7. Ding, H.; Han, Z.; Li, J.; Li, X.; Dong, Z.; Zhao, J.; Wang, S.; Shao, T. Effect of Fibrolytic Enzymes, Cellulolytic Fungi and Lactic Acid Bacteria on Fermentation Characteristics, Structural Carbohydrate Composition and In Vitro Digestibility of Rice Straw Silage. *Fermentation* **2022**, *8*, 709. [CrossRef]
8. Yi, Q.; Yu, M.; Wang, P.; Du, J.; Zhao, T.; Jin, Y.; Tang, H.; Yuan, B. Effects of Moisture Content and Silage Starter on the Fermentation Quality and In Vitro Digestibility of Waxy Corn Processing Byproduct Silage. *Fermentation* **2023**, *9*, 1025. [CrossRef]
9. Bunyamin, A.; Hıdır, G. Effects of lactic acid bacteria inoculant on quality, fermentation profile and nutritive value of alfalfa silage at different ensiling period. *Ank. Univ. Vet. Fak. Derg.* **2020**, *67*, 281–287. [CrossRef]
10. Usman, S.; Li, F.; An, D.; Shou, N.; Deng, J.; Zhang, Y.; Guo, X.; Shen, Y. Lignocellulose Degradation and Enzymatic Hydrolysis of Soybean Incorporated Sorghum Silage Inoculated with Feruloyl-Esterase Producing Lactobacillus plantarum. *Fermentation* **2022**, *8*, 70. [CrossRef]
11. Lin, S.; Yanlin, X.; Yanzi, X.; Rigele, T.; Xiaoguang, W.; Na, N.; Nanping, W.; Moge, Q.; Yi, Z.; Yimin, C. Community Synergy of Lactic Acid Bacteria and Cleaner Fermentation of Oat Silage Prepared with a Multispecies Microbial Inoculant. *Microbiol. Spectr.* **2023**, *11*, e00705-23. [CrossRef]
12. Yan, Y.; Li, X.; Guan, H.; Huang, L.; Ma, X.; Peng, Y.; Li, Z.; Nie, G.; Zhou, J.; Yang, W. Microbial community and fermentation characteristic of Italian ryegrass silage prepared with corn stover and lactic acid bacteria. *Bioresour. Technol.* **2019**, *279*, 166–173. [CrossRef] [PubMed]
13. Guo, L.; Wang, X.; Lin, Y.; Yang, X.; Ni, K.; Yang, F. Microorganisms that are critical for the fermentation quality of paper mulberry silage. *Food Energy Secur.* **2021**, *10*, e304. [CrossRef]
14. Zhao, S.S.; Wang, Y.P.; Yang, F.Y.; Wang, Y.; Zhang, H. Screening a *Lactobacillus* plantarum strain for good adaption in alfalfa ensiling and demonstrating its improvement of alfalfa silage quality. *J. Appl. Microbiol.* **2020**, *129*, 233–242. [CrossRef] [PubMed]
15. Hartutik, H.; Marjuki, M.; Huda, A.N.; Ndaru, H.P.; Arsianty, Y.N.; Rohmah, B.N. The Use of Molasses as Additive with Different Ensiling Time and Physical Quality, pH and Nutritive of Value Maize Stover Silage. *J. Nutr. Ternak Trop.* **2021**, *4*, 88–92. [CrossRef]
16. Broderick, G.A.; Jinhe, K. Automated Simultaneous Determination of Ammonia and Total Amino Acids in Ruminal Fluid and In Vitro Media. *J. Dairy. Sci.* **1980**, *63*, 64–75. [CrossRef]
17. Gheller, L.S.; Ghizzi, L.G.; Takiya, C.S.; Grigoletto, N.T.S.; Silva, T.B.P.; Marques, J.A.; Dias, M.S.S.; Freu, G.; Rennó, F.P. Different organic acid preparations on fermentation and microbiological profile, chemical composition, and aerobic stability of whole-plant corn silage. *Anim. Feed. Sci. Technol.* **2021**, *281*, 115083. [CrossRef]
18. Chen, R.; Li, M.; Yang, J.; Chen, L.; Zi, X.; Zhou, H.; Tang, J. Exploring the effect of wilting on fermentation profiles and microbial community structure during ensiling and air exposure of king grass silage. *Front. Microbiol.* **2022**, *13*, 971426. [CrossRef]
19. Thomas, T.A. An automated procedure for the determination of soluble carbohydrates in herbage. *J. Sci. Food Agric.* **1977**, *28*, 639–642. [CrossRef]

20. Soest, P.J.V.; James, B.R.; Lewis, B.A. Methods for Dietary Fiber, Neutral Detergent Fiber, and Nonstarch Polysaccharides in Relation to Animal Nutrition. *J. Dairy Sci.* **1991**, *74*, 3583–3597. [CrossRef]
21. *DIN EN ISO 16634-1:2009-07*; Food Products_-Determination of the Total Nitrogen Content Using the Dumas Combustion Method and Calculation of the Crude Protein Content_-Part_1: Oilseed Meals and Animal Feed (ISO_16634-1:2008). German Version EN_ISO_16634-1:2008; ISO: Geneva, Switzerland, 2019. [CrossRef]
22. Chunshuang, L.; Dixian, Z.; Wenjie, M.; Guowei, Y.; Aijie, W.; Qilin, W.; Dong Hoon, L. Denitrifying sulfide removal process on high-salinity wastewaters in the presence of Halomonas sp. *Appl. Microbiol. Biotechnol.* **2015**, *100*, 1421–1426. [CrossRef]
23. Chen, S.; Zhou, Y.; Chen, Y.; Gu, J. fastp: An ultra-fast all-in-one FASTQ preprocessor. *Bioinformatics* **2018**, *34*, i884–i890. [CrossRef] [PubMed]
24. Magoc, T.; Salzberg, S.L. FLASH: Fast length adjustment of short reads to improve genome assemblies. *Bioinformatics* **2011**, *27*, 2957–2963. [CrossRef] [PubMed]
25. Bolyen, E.; Rideout, J.R.; Dillon, M.R.; Bokulich, N.A.; Abnet, C.C.; Al-Ghalith, G.A.; Alexander, H.; Alm, E.J.; Arumugam, M.; Asnicar, F.; et al. Reproducible, interactive, scalable and extensible microbiome data science using QIIME 2. *Nat. Biotechnol.* **2019**, *37*, 852–857. [CrossRef] [PubMed]
26. Douglas, G.M.; Maffei, V.J.; Zaneveld, J.R.; Yurgel, S.N.; Brown, J.R.; Taylor, C.M.; Huttenhower, C.; Langille, M.G.I. PICRUSt2 for prediction of metagenome functions. *Nat. Biotechnol.* **2020**, *38*, 685–688. [CrossRef] [PubMed]
27. Patrick, D.S.; Sarah, L.W.; Thomas, R.; Justine, R.H.; Martin, H.; Emily, B.H.; Ryan, A.L.; Brian, B.O.; Donovan, H.P.; Courtney, J.R. Introducing mothur: Open-Source, Platform-Independent, Community-Supported Software for Describing and Comparing Microbial Communities. *Appl. Environ. Microbiol.* **2009**, *75*, 7537–7541. [CrossRef]
28. Nicola, S.; Jacques, I.; Levi, W.; Dirk, G.; Larisa, M.; Wendy, S.G.; Curtis, H. Metagenomic biomarker discovery and explanation. *Genome Biol.* **2011**, *12*, R60. [CrossRef]
29. Dai, H.-l.; Xiaoli, T.; Weian, D.; Jianping, W. Effects of silage additives on nutritional quality and silage quality of triticale and rye. *Acta Prataculutral Sci.* **2019**, *28*, 211. [CrossRef]
30. Jiang, H.; Wang, H.; Bao, B.; Qu, H.; Wang, J.; Sun, L.; Liu, B.; Gao, F. Effect of Compound Additives on Nutritional Composition, Fermentation Quality, and Bacterial Community of High-Moisture Alfalfa Silage. *Fermentation* **2023**, *9*, 453. [CrossRef]
31. Liu, X.; Wang, A.; Zhu, L.; Guo, W.; Guo, X.; Zhu, B.; Yang, M. Effect of additive cellulase on fermentation quality of whole-plant corn silage ensiling by a Bacillus inoculant and dynamic microbial community analysis. *Front. Microbiol.* **2023**, *14*, 1330538. [CrossRef]
32. Bai, J.; Ding, Z.; Ke, W.; Xu, D.; Wang, M.; Huang, W.; Zhang, Y.; Liu, F.; Guo, X. Different lactic acid bacteria and their combinations regulated the fermentation process of ensiled alfalfa: Ensiling characteristics, dynamics of bacterial community and their functional shifts. *Microb. Biotechnol.* **2021**, *14*, 1171–1182. [CrossRef] [PubMed]
33. Muhammad, R.; Senji, U. The effect of cellulase addition on nutritional and fermentation quality of barley straw silage. *Asian-Australas. J. Anim. Sci.* **1993**, *6*, 383–388. [CrossRef]
34. Ling, W.; Zhang, L.; Feng, Q.; Degen, A.A.; Li, J.; Qi, Y.; Li, Y.; Zhou, Y.; Liu, Y.; Yang, F. Effects of Different Additives on Fermentation Quality, Microbial Communities, and Rumen Degradation of Alfalfa Silage. *Fermentation* **2022**, *8*, 660. [CrossRef]
35. Zhao, C.; Wang, L.; Ma, G.; Jiang, X.; Yang, J.; Lv, J.; Zhang, Y. Cellulase Interacts with Lactic Acid Bacteria to Affect Fermentation Quality, Microbial Community, and Ruminal Degradability in Mixed Silage of Soybean Residue and Corn Stover. *Animals* **2021**, *11*, 334. [CrossRef] [PubMed]
36. Li, S.; Wang, H.; Luo, M.; Wu, B.; Duan, H.; Ma, X.; Ren, J. Effects of cellulase and xylanase additives on fermentation quality and nutrient composition of silage maize. *Circ. Agric. Syst.* **2023**, *3*, 8. [CrossRef]
37. Tatli Seven, P.; Seven, I.; Iflazoglu Mutlu, S.; Nur Yildirim, E. Silage Additives Usage in Improving Fermentation Quality of Alfalfa Silage: A Review. *Anim. Vet. Sci.* **2021**, *9*, 175–180. [CrossRef]
38. Hartutik; Dewangga, R.N. The Effect of Various Additive with Different Level on pH, Total Lactic Acid and Total Lactic Acid Bacteria Maize Stover (Zea Mays L.) Silage. In Proceedings of the 3rd International Conference on Environmentally Sustainable Animal Industry 2022 (ICESAI 2022), Malang, Indonesia, 13 October 2022; pp. 341–348.
39. Du, Z.; Yamasaki, S.; Oya, T.; Cai, Y. Cellulase-lactic acid bacteria synergy action regulates silage fermentation of woody plant. *Biotechnol. Biofuels Bioprod.* **2023**, *16*, 125. [CrossRef] [PubMed]
40. Silva, V.P.; Pereira, O.G.; Leandro, E.S.; Paula, R.A.; Agarussi, M.C.N.; Ribeiro, K.G. Selection of Lactic Acid Bacteria from Alfalfa Silage and Its Effects as Inoculant on Silage Fermentation. *Agriculture* **2020**, *10*, 518. [CrossRef]
41. Deng, X.; Jia, Y.; Ge, G.; Wang, Z.; Liu, M.; Bao, J.; Zhao, M.; Si, Q.; Liu, Y.; Zhao, W. Microbiomics and volatile metabolomics-based investigation of changes in quality and flavor of oat (Avena sativa L.) silage at different stages. *Front. Plant Sci.* **2023**, *14*, 1278715. [CrossRef] [PubMed]

42. Zhao, G.; Wu, H.; Li, L.; He, J.; Hu, Z.; Yang, X.; Xie, X. Effects of applying cellulase and starch on the fermentation characteristics and microbial communities of Napier grass (Pennisetum purpureum Schum.) silage. *J. Anim. Sci. Technol.* **2021**, *63*, 1301–1313. [CrossRef]
43. Zhang, Z.; Zhao, K.; Yang, S.; Min, L.; Tong, X.; Chen, W.; Li, D. Analysis on Fermentation Quality, Chemical Composition and Bacterial Communities of Corn Straw and Soybean Straw Mixed Silage. *Fermentation* **2023**, *9*, 500. [CrossRef]
44. He, L.; Zhou, W.; Wang, C.; Yang, F.; Chen, X.; Zhang, Q. Effect of cellulase and *Lactobacillus* casei on ensiling characteristics, chemical composition, antioxidant activity, and digestibility of mulberry leaf silage. *J. Dairy. Sci.* **2019**, *102*, 9919–9931. [CrossRef] [PubMed]
45. Liu, W.; Si, Q.; Sun, L.; Wang, Z.; Liu, M.; Du, S.; Ge, G.; Jia, Y. Effects of Cellulase and Xylanase Addition on Fermentation Quality, Aerobic Stability, and Bacteria Composition of Low Water-Soluble Carbohydrates Oat Silage. *Fermentation* **2023**, *9*, 638. [CrossRef]
46. Chapla, D.; Pandit, P.; Shah, A. Production of xylooligosaccharides from corncob xylan by fungal xylanase and their utilization by probiotics. *Bioresour. Technol.* **2012**, *115*, 215–221. [CrossRef] [PubMed]
47. Xiaolin, W.; Jiamei, S.; Zihan, L.; Guangning, Z.; Yonggen, Z. Fermentation Quality and Microbial Community of Corn Stover or Rice Straw Silage Mixed with Soybean Curd Residue. *Animals* **2022**, *12*, 919. [CrossRef] [PubMed]
48. Agustinho, B.C.; Daniel, J.L.P.; Zeoula, L.M.; Alcalde, C.R.; Machado, E.; Bragatto, J.M.; Schneider, C.R.; Santos, N.W.; Matumoto-Pintro, P.T.; Saraiva, B.R.; et al. Enzymatic effects of Pleurotus ostreatus spent substrate on whole-plant corn silage and performance of lactating goats. *J. Dairy. Sci.* **2021**, *104*, 11660–11672. [CrossRef]

MDPI AG
Grosspeteranlage 5
4052 Basel
Switzerland
Tel.: +41 61 683 77 34

Fermentation Editorial Office
E-mail: fermentation@mdpi.com
www.mdpi.com/journal/fermentation

www.ingramcontent.com/pod-product-compliance
Lightning Source LLC
Chambersburg PA
CBHW062032220926
51850CB00010B/2525